RAPID PROTOTYPING

RAPID PROTOTYPING

Principles and Applications

RAFIQ NOORANI, Ph.D.
Professor of Mechanical Engineering
Loyola Marymount University
Los Angeles, CA

WILEY

JOHN WILEY & SONS, INC.

Library of Congress Cataloging-in-Publication Data:

Noorani, Rafiq.
   Rapid prototyping : principles and applications/Rafiq Noorani.
      p. cm.
   Includes bibliographical references and index.
   ISBN-13: 978-0-471-73001-9
   ISBN-10: 0-471-73001-7 (cloth)
      1. Rapid prototyping.   I. Title.
   TS155.6.N66 2005
   620'.0042—dc22

                                                           2005004698

Printed in the United States of America

10 9 8 7 6 5 4 3 2 1

CONTENTS

PREFACE

The role of engineers, particularly those engaged in design and manufacturing of competitive products, has been changing over the last several years due to the emergence of a powerful and sophisticated tool known as *rapid prototyping*. Rapid prototyping (RP), also called layered manufacturing or solid free-form fabrication, refers to the physical modeling of a new design concept early in the product design cycle taken directly from computer-aided design (CAD) data without the use of tooling. Some of the benefits attributed to rapid prototyping are (1) reduced lead time to produce prototypes, (2) improved ability to visualize the part due to its physical existence, (3) early detection and correction of design errors, and (4) increased capability to compute mass properties of components and assemblies.

An impressive number of educational institutions, manufacturing organizations, and industries have discovered rapid prototyping and its potential benefits. Many colleges and universities have purchased and implemented rapid prototypes into their curricula. Students are routinely taught the principles and applications of rapid prototyping and use RP tools in their laboratory and design/capstone projects. Automobile manufacturers and suppliers are using RP to prototype various mechanical components. Aerospace companies use RP to produce parts and assemblies for both commercial and military aircraft, as well as in the product development cycle. Manufacturers of business machines use RP to prototype computer parts, copier, telephone, and fax machine parts. The growth of the RP market has been explosive. Since its inception in 1988, the market has grown steadily (58% a year on average).

The primary purpose of this text is to present the principles and applications of rapid prototyping as it relates to the world of design and manufacturing. The text is designed for a junior- or senior-level course in manufacturing automation. The upper-level undergraduate students of a mechanical, industrial, manufacturing, or aerospace engineering curriculum are ideal candidates for a course such as this. The text can also be used as a supplement to manufacturing or engineering development texts that are currently being used. Although the primary audience of this book is the undergraduate student of an engineering discipline, it is also intended to serve as a reference text for industrial and academic practitioners interested in competitive product design and manufacturing.

The book is organized into 11 chapters. The goals of each chapter follow:

- *Chapter 1 Introduction*   This chapter introduces the readers to the world of rapid prototyping. Definition, historical development, important developments, and applications of RP are described.

- *Chapter 2 Principles of Rapid Prototyping*   This chapter provides an overview of the automated fabrication process and where RP fits into this process. It also describes the basic steps in making prototypes, that is, solid model creation, data conversion, part fabrication, and software considerations.

- *Chapter 3 Liquid-Based RP Systems*   This chapter provides an overview of the various types of liquid-based RP systems that are commercially available to university and industry.

- *Chapter 4 Solid-Based RP Systems*   This chapter provides an overview of the various types of solid-based RP systems that are commercially available to university and industry.

- *Chapter 5 Powder-Based RP Systems*   This chapter provides an overview of the various types of powder-based RP systems that are commercially available to university and industry.

- *Chapter 6 Materials for Rapid Prototyping*   RP materials play a key role in the quality, strength, and accuracy of prototypes. This chapter describes properties of various materials such as photopolymers, plastics, powder, composites, and the like.

- *Chapter 7 Reverse Engineering*   Reverse engineering (RE) refers to the generation of accurate surface models from physical prototypes. RE plays an important role in shortening product lead time. This chapter discusses the background, scanning techniques, data conversion systems, and future trends in RE.

- *Chapter 8 Rapid Tooling*   Rapid tooling produced from RP methods allows for faster product development than existing approaches. This chapter describes soft and hard tooling, metal casting, Quickcast, mold and die making, and other tooling techniques.

- *Chapter 9 Medical Applications in Rapid Prototyping*   This chapter compiles information on some current medical applications of rapid prototyping technology. It will also familiarize the readers with imaging technologies including new computer software packages that interface medical practices with rapid prototyping techniques. The materials for medical RP and other applications such as anthropology and biochemistry are also discussed.

- *Chapter 10 Industry Perspectives* This chapter describes guidelines for implementing RP in the workplace, operations and management issues, and service bureaus. It concludes with information about various RP organizations/consortia that are valuable resources for RP users and developers.

- *Chapter 11 Research and Development* The success of RP depends on the continued research and development in the various engineering and manufacturing fields. These challenges include improving the RP processes as well as improving part accuracy, build speed, and material properties. This chapter presents examples of research and development efforts of the author and his students at Loyola Marymount University by way of illustration of the type of research that is both useful to industry and instructive for students.

Although this text should be considered an overview of RP technology, enough specific information and worthwhile exercises are presented to give the student an excellent working knowledge of the technology. Since RP spans several technologies, for example, computers, polymer science, lasers, and robotics, an exhaustive presentation of the theory involved would be well outside the scope of this text. A complete understanding of the individual technologies that have been combined to make rapid prototyping possible is not necessary to be proficient in the application of RP, albeit useful for more in-depth study or research into new applications or materials.

From the standpoint of organizing an undergraduate course in rapid prototyping as described earlier, the instructor should consider the text in three sections. Chapters 1 through 5 represent the most basic information necessary to understand layered manufacturing processes in general and to be conversant in the nomenclature and most widely recognized acronyms. A significant amount of time, including numerous review exercises, should be set aside to ensure the student has adequate understanding of this essential material.

Chapters 7 through 9 should be considered the "applications" section of the text. The information presented in these chapters should stimulate both discussion and outside research on the ever-expanding arena of RP applications. These first two sections should be considered a minimum exposure to the subject and can be adequately covered in one semester. A class project done in parallel with the presentation of the second section, along with field trips to active users of the technology, will also serve to cement the concepts in the students' mind.

Chapters 6, 10, and 11 can be given brief coverage if time does not permit detailed study. Issues related to materials will normally be given sufficient coverage while presenting Chapters 3 through 5. Chapters 10 and 11 are presented as a means of providing access to information recommended for potential users of the RP technology in a university or industrial setting. Although these chapters are not required for the beginning student, they can provide valuable information for further study and application of the technology to "real-world" problems.

ACKNOWLEDGMENTS

I have been very fortunate in having many talented colleagues, friends, and students during the 25 years in which I have been actively involved in computer-aided design and computer-aided manufacturing (CAD/CAM). Many ideas in this book have originated from close friendships, co-authorships, collaborative projects, and formal and informal discussions at various conferences.

Looking back at my career in computer-aided design and manufacturing, I remember my work as the graduate program coordinator at the University of Louisiana at Lafayette with Dr. Stan Smith who supported me and encouraged my involvement in research in design and manufacturing. I am grateful to Dr. M. A. Aziz of Gonzaga University who gave me an opportunity to work in the Mechanical Engineering Department and encouraged me to write my first successful National Science Foundation (NSF) proposal on robotics and control.

I am grateful to the faculty of the Mechanical Engineering Department of Loyola Marymount University (LMU), which brought me to LMU in 1989 to develop the CAD/CAM program. At LMU, I am thankful to Dr. Omar Es-Said who selected me to be the co-principal investigator of the NSF Research Experiences for the Undergraduates (REU) program. I have had a satisfying experience of conducting research with the undergraduate students at LMU and the local community college. Some exciting results of the REU program are presented in Chapter 11. I am also thankful to Dr. Mel Mendelson of the department with whom I have worked on many university and industry projects on rapid prototyping and design of experiments. Dr. Mendelson also wrote Chapter 6 of this book. I am also grateful to Dr. Christof Stotko of

EOS, Germany who helped me write the section on Laser Sintering as a means to e-manufacturing.

Collaboration with Northrop-Grumman Corporation in El Segundo, California, has been a great source of support and encouragement for my experience in rapid prototyping. I am pleased to acknowledge the support and contribution of Mr. Boris Fritz of Northrop-Grumman toward the development of the book. I am also grateful to Mr. Richard Zoia who teaches rapid prototyping at UCLA as a part-time instructor and who thoroughly checked the contents of the book and gave me invaluable comments and suggestions for the improvement of the text.

Support from the National Science Foundation has been invaluable in developing the CAD/CAM Laboratory and the Rapid Prototyping Laboratory at LMU. It is the NSF grant that I received in 1995 that helped me buy the first rapid prototyping (RP) machine and start my work on RP. I am grateful to Parsons Foundation who gave me a grant in 1997 to purchase my second RP machine as well as a coordinate measurement machine (CMM).

As an international professor, I have worked very closely with Dr. Toshihiro Ioi of Chiba Institute of Technology (CIT) in Japan, Dr. Kwang Sun Kim of Korea University of Technology and Education (KUT) in Cheonan, and Dr. Kwan Heng Lee of Kwangju Institute of Science and Technology (KJIST in Kwangju in South Korea). Dr. Lee of KJIST developed Chapter 7 (Reverse Engineering) for the book. These professors have invited me to give lectures on rapid prototyping for their classes and for local industries and for classes at vocational and technical colleges. I am very grateful to these gentlemen for their continued support and encouragement.

I am very grateful to my anonymous reviewers who made corrections and improvements to the chapter contents. I am thankful to my students Mathew Graham, Joey Weinman, Chris Lapiratanagool, Daniel Horvath, and others who worked with me very closely to develop the chapters of this book. I thank my daughter Sabrina Noorani who worked with me very closely to write Chapter 9 (Medical Applications of RP). I must acknowledge the encouragement I have received from my wife Zarina Noorani to complete the book.

And finally, I would like to thank John Wiley & Sons. I am very thankful to Bob Argentieri, my acquisition editor, Millie Torres, senior production editor, and Lindsay Orman, executive assistant at John Wiley, who worked with me very closely to convert the manuscript into a textbook. As the publisher, John Wiley turned this book into a reality.

ABOUT THE AUTHOR

Rafiq Noorani received his B.Sc. in mechanical engineering from Bangladesh University of Engineering and Technology (BUET), an M. Engr. in Nuclear Engineering, and a Ph.D. in mechanical engineering from Texas A&M University.

He is currently a tenured Professor of Mechanical Engineering at Loyola Marymount University of Los Angeles, California. He previously taught engineering at Texas A&M University, the University of Louisiana at Lafayette, and Gonzaga University in Spokane, Washington, before joining Loyola Marymount University. His teaching interests are in CAD/CAM, rapid prototyping, and nanotechnology. He has received over 33 external grants totaling over $2 million including 6 grants from the National Science Foundation. He has published 9 journal papers and 50 conference papers in national and international journals and conferences. With the help of a recent NSF grant, he is currently working with his colleagues to develop an undergraduate program in nanotechnology at LMU.

As an international professor, he teaches and lectures on CAD/CAM, rapid prototyping, and nanotechnology in many countries including Japan, South Korea, and Bangladesh. Dr. Noorani is a professional member of the American Society of Mechanical Engineers (ASME), the Society of Manufacturing Engineers (SME), and the American Society for Engineering Education (ASEE).

Dedication
This book is dedicated to my late father, Osman Ali Noorani,
a devoted teacher and mentor who touched the lives of many students
at Kalaroa High School and beyond for more than 50 years and who,
to this day, continues to be my personal guide and inspiration.

1

INTRODUCTION

Mastering the art of rapidly prototyping components, subassemblies, and products is vital for any product development process. Over the last few years, new and exciting technologies have emerged that are changing the ways that products have been launched. *Rapid prototyping* (RP) is such a technology and has revolutionized the design and manufacturing of new products. This technology is being used to fabricate physical solid models for early verification of concepts, for example, form, fit, and function, as well as reducing lead times for product development.

In this opening chapter, we shall define rapid prototyping, study its history, and describe the importance of this technology for product development. Applications of this technology through case study and future trends will also be discussed.

1.1 INTRODUCTION

The keys to regaining competitiveness in most design and manufacturing industries are quality, productivity, reduced costs, customer satisfaction, and responsiveness in bringing new products to the marketplace [1–3]. One of the methods for attaining these attributes is rapid prototyping. Rapid prototyping takes information from a three-dimensional (3D) computer-aided design (CAD) database and produces a solid model (prototype) of the design [4–8]. One can turn a design concept into a solid prototype and test it for form, fit, and function at a fraction of the cost and time of traditional prototyping methods [4, 5]. The benefits of RP include (1) reduced lead times to produce prototyped components, (2) improved ability to visualize the part geometry due to its physical existence, (3) earlier detection and reduction of design errors, (4) increased capability to compute mass properties of components and assemblies [6]. RP is advantageous in the elimination of waste and costly late design changes.

The traditional method of prototyping a part is machining, which can require significant lead times from several days to many weeks, depending on the complexity of the part and availability of materials, equipment, and labor [5]. A number of RP technologies [3–5] such as stereolithography, selective laser sintering, fused deposition modeling, and 3D printing are available for producing plastic parts directly from design data created in any one of many commercially available CAD systems.

In addition to making prototypes, RP techniques are now being used to produce tooling (known as rapid tooling). Rapid tooling is helping industries eliminate the costly and time-consuming practice of machining metal tools and dies. Another natural extension of RP is rapid manufacturing—the automated production of final products directly from CAD files. At present, only a few companies are capable of producing salable products, but the number will increase as technologies, processes, and material properties continue to improve. Rapid manufacturing may never eliminate high-speed, high-volume manufacturing processes, in which large production runs are more economical, but it can make the development process more efficient.

In this opening chapter, we introduce the reader to the world of rapid prototyping. The definition, historical development, important recent developments, applications of RP systems, and the future of RP will be discussed.

1.2 WORLD OF RP

The field of rapid prototyping encompasses a wide variety of new methods, technologies, and applications that have already stimulated some fascinating research. Many companies have found exciting new ways to improve product development processes and enhance profitability. The prospect of being a part of this new technology with its promise of radical improvement in the way business is done should be as highly motivating to the reader as it is the author. In this section we describe some definitions, applications of RP, and give examples of some companies that are using RP.

1.2.1 What Is RP?

Rapid prototyping, as described in this text, refers to the fabrication of a physical, three-dimensional part of arbitrary shape directly from a numerical description (typically a CAD model) by a quick, totally automated, and highly flexible process. According to *Wohler's Report 2000,* RP is defined as: a special class of machine technology that quickly produces models and prototype parts from 3-D data using an additive approach to form the physical models. These two definitions are the "classic" definitions that describe the systems and technologies that created the rapid prototyping industries in the late 1980s.

As we have already mentioned, RP is a new technology that is having a profound effect on the product development process of design and manufacturing industries worldwide. It is called rapid prototyping because it can prototype parts very rapidly in most cases, that is, in hours rather than in days or weeks. This technology is quickly expanding to include rapid tooling and rapid manufacturing. So, *rapid prototyping, tooling, and manufacturing* (RPTM) is a term that is also widely used these days [7]. Other less commonly used terms include *desktop manufacturing, direct CAD manufacturing,* and *instant manufacturing.* The rationale behind the use of these terms is also the speed, ease of use, and the desktop nature of some RP equipment.

The unique characteristic of RP is that it makes prototypes one layer at a time or layer by layer. Another group of terms that emphasizes this characteristic includes *layered manufacturing, material deposit manufacturing,* and *material addition manufacturing.*

The last group of terms for RP emphasizes the words "solid," "freeform," and "fabrication." This group includes *solid freeform fabrication* and *solid freeform manufacturing.* The word "solid" refers to the final solid state of the material, although the initial state may be solid, liquid, or powder. The word "freeform" stresses the fact that RP can prototype complex shapes with little or no constraint on its form. RP is also related to "automated fabrication," which describes new technologies for generating 3D objects from computer files.

Applications of RP   There are many types and classes of physical prototypes, but their main purpose is to minimize risk during the product development process. Some of the specific applications of RP follow:

- Communication of product characteristics
- Engineering concept definition
- Form, fit, and function testing
- Engineering change clarification
- Client presentations and consumer evaluations
- Bid proposals and regulation certification
- Styling, ergonomic studies
- Facilitate meeting schedule and making milestones
- Masters for silicone rubber tooling
- Masters for spray metal tooling (all processes)
- Masters for epoxy tooling to be used for injection molding
- Master/pattern for investment casting
- Tooling for injection molding

Industrial Uses of RP Rapid prototyping has found widespread acceptance by companies that provide products to consumers. Many companies, small and big, use RP to enhance their design and manufacturing processes. Some of the major companies currently using RP technology follow:

General Motors	Pratt & Whitney	Chrysler
Motorola	Texas Instruments	General Motors
AT&T	Biomet	TRW
Apple Computer	Lockheed Martin	Boeing
Eastman Kodak	3M	Eureka
J&J Orthopaedic	AMP	Mercedes Benz
GE Aircraft Engines	ITT Defense	Hasbro
Rockwell	Mattel	Gillette
Ford	Northrop Grumman	Baxter

1.2.2 Basic Process

Many different processes are available today, using a variety of materials, such as wax, plastic, and metals with a variety of processes, such as stereolithography and fused deposition modeling. However, almost all the processes operate by forming solid objects, using either liquid or powder-based materials. The basic process of RP consists of the following steps [7, 8]:

1. Create a CAD model of the design.
2. Convert the CAD model to stereolithography (STL) file format.
3. Slice the STL file into two-dimensional (2D) cross-sectional layers.
4. Grow the prototype.
5. Clean and finish the model.

The first step is to create the electronic solid model using standard drafting software packages, such as AutoCAD, Pro/Engineer, or Solid Works, to name a few.

The second step involves the conversion of the drawing file of the CAD model into an STL format. Since various CAD software packages use different algorithms to represent solid objects, STL file format has been selected as the de facto standard in the rapid prototyping industry. The STL file represents a three-dimensional surface of an assembly as planar triangles. The file contains the coordinates of the vertices and the direction of the outward normal of each triangle. The STL file format is the best file format to represent all surfaces, in preparation for the "slicing" algorithms. More about the STL file format will be studied in Chapter 2.

The third step involves slicing the STL file using a proprietary software program, provided by the manufacturer of the RP machine in which the model is to be produced. The preprocessing software imports the STL file and lets the user orient the part and adjust the size and slice thickness of the model.

The slice thickness may vary from 0.01 to 0.7 mm depending on the capabilities of the RP machine. Lower slice thickness increases the accuracy of the prototype but also increases the time to build the model. The preprocessing software may also generate a structure to support portions of the model during its buildup. Supports are necessary for creating features such as overhangs, internal cavities, and thin-walled sections. The RP-processing software also provides information about how much time and material will be required to make the prototype.

The fourth step involves the actual making of the prototype. Once the STL file is processed and saved, it is sent to the RP machine. At this time, the RP machine acts as a printer. Building the prototype one layer at a time. Most modern RP machines can operate unattended once the initial setup is completed.

The final step in prototyping is removing the part from the machine and cleaning it before use. This step is also known as postprocessing. It also involves postcuring of photosensitive materials, sintering powder materials, and removing the support materials. Some prototypes are also subjected to surface treatment, such as sanding, sealing, or painting to improve their appearance and durability.

1.2.3 Industries Using RP

Design and manufacturing industries use RP to reduce the time to market their products and to cut manufacturing costs. The early uses of prototypes were mainly for visualization, to check form, fit, and function, and the early verification of design error. Prototypes are now used for many other purposes. Functional models and fit/assembly represent about 36% of all RP models used. About 25% of RP models are used for patterns for prototype tooling and metal casting. This area of application is expected to increase dramatically in the near future. Figure 1.1 shows how companies are using RP models [9]. Figure 1.2 shows the same data as Figure 1.1; however, the length of each bar reflects the numerical responses received from those surveyed.

1.2.4 Growth of RP

While the RP industry continues to grow, the rate of growth has been flat for the last few years due to the lethargy in the overall economy. In spite of this, over the last few years, established RP companies released new technologies, new materials, and new applications. Figure 1.3 shows the growth of rapid prototyping since 1988. Sales for 2004 and 2005 are forecasts.

Increase in RP Models Worldwide RP users are producing more and more RP models. In 2001, 3.55 million models or prototypes were produced, up 18.3% from 3 million units made in 2000, according to *Wohler's Report 2004.* The growth of RP models is charted in Figure 1.4.

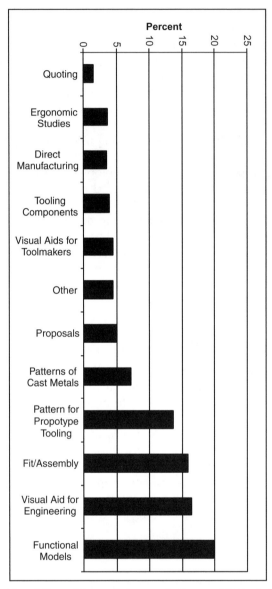

Figure 1.1 How RP models are being used. *Source:* Wohler's Report 2004.

Sales of RPs by Regions   Although RP is a worldwide phenomenon, U.S. manufacturers account for 80% of the unit sales, followed by Japan, Europe, and others. Figure 1.5 shows the total systems sold by region [9].

Units Sold by U.S. Companies   Among the U.S. manufacturers of RP equipment, 3D Systems has been the leader in the industry since the beginning. Since its recent acquisition of DTM System, it became an even more dominant

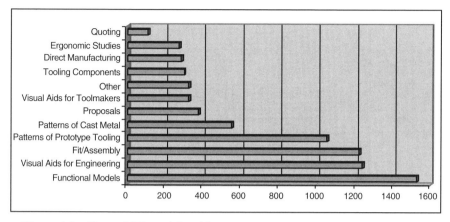

Figure 1.2 Uses of RP models with responses. *Source:* Wohler's Report 2004.

company in the United States. With combined sales of 415 units sold in 2001, 3D Systems controls 40% of the unit sales in the United States, followed by Stratasys, which sold 277 units. Figure 1.6 shows the units sold by U.S. companies. For cost-conscious companies, all major RP companies have developed low-cost RP machines such as ThermoJet (3D Systems), Z400 (Z Corp.), and Dimension (Stratasys). The low-cost machines are like concept modelers that are used for concept verification, form, fit, and function, and

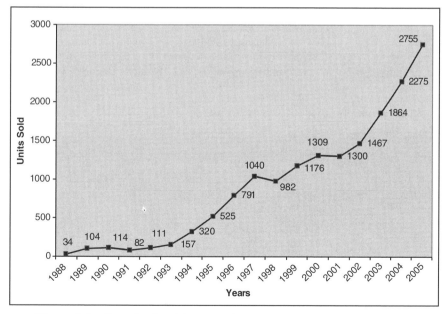

Figure 1.3 Growth of rapid prototyping. *Source:* Wohler's Report 2004.

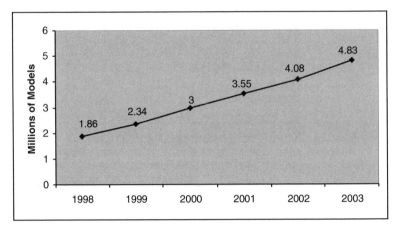

Figure 1.4 Growth of RP models. *Source:* Wohler's Report 2004.

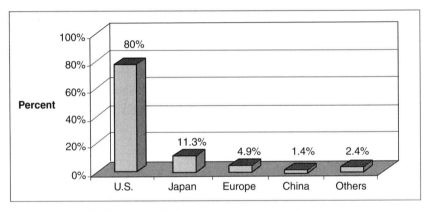

Figure 1.5 Systems sold by region. *Source:* Wohler's Report 2004.

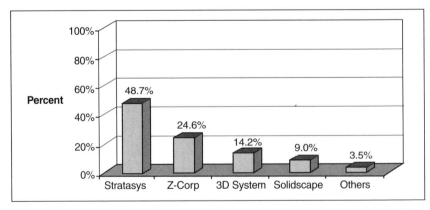

Figure 1.6 Units sold by U.S. companies. *Source:* Wohler's Report 2004.

early verification of design errors and are compatible with an office environment. According to *Wohler's Report 2002,* the low-priced systems are outselling the expensive, high-end models of the companies.

1.2.5 Installations by Countries

Rapid prototyping is a technology that is used by most industrialized countries of the world. The general goals, objectives, and needs for rapid prototyping are the same in the United States, Europe, and Japan, although the emphases may be different for different countries. In Japan, the emphasis on accuracy is the predominant consideration for using RP, while in Europe there is more emphasis on developing prototypes in metal components and tooling. As mentioned before, the RP technology was developed in the United States, and the United States is still the leader of this technology, both in terms of its development and applications [9]. A detailed breakdown of unit sales by country can be found in *Wohler's Report 2004.*

1.2.6 Technology Development

Rapid prototyping is not a technology by itself; rather, it draws on other technologies such as CAD, computer-aided manufacturing (CAM), and computer numerical control (CNC). Figure 1.7 traces the historical developments of various technologies from the estimated date of inception.

1.3 HISTORY OF RP*

For the most part, RP is a technology that was launched during the 1980s. While the modern concept of RP was conceived, developed, and promoted in the 1980s, the origin of this technology dates back to the 1890s. Beaman et al. [10] have studied the history of RP extensively from its roots to today's

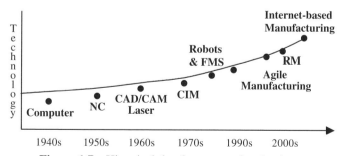

Figure 1.7 Historical developments of technology.

applications. In this section, the historical development of RP will be described.

1.3.1 Early History

Rapid prototyping is also called *layered manufacturing*. The concept of building something layer by layer can be traced back to at least two technical areas: *topography* and *photosculpture* [10].

Topography It had been reported as early as 1890 that Blanther [11] suggested a layered method for making a mold for topographical maps. The method consisted of impressing topographical contour lines on a series of wax plates, cutting the wax plates on the contour lines, and then stacking and smoothing the wax section. This method was used to produce both positive and negative three-dimensional surfaces corresponding to the terrain indicated by the contour lines. After suitably backing these surfaces, a printed paper map was then pressed between the positive and negative forms to create a raised relief map. An example is shown in Figure 1.8. Work on similar concepts have been reported by Perera [12], Zang [13], and Gaskin [14]. In 1972, Matsubara [15] proposed a topographical process using photohardening materials in which a photopolymer resin is coated onto refractory particles that are then spread into a layer and heated to form a coherent sheet. Light is then selectively projected or scanned onto this sheet to harden a defined portion

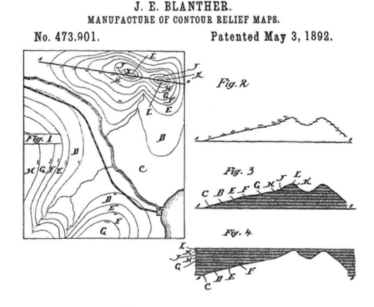

Figure 1.8 Layered map concept developed by Blanther [11].

of it. A solvent is used to dissolve any unscanned, unhardened areas. The thin layers formed this way are then stacked together to form a casting mold. DiMatteo [16] realized the power of this stacking technique and thought that it could be used for producing surfaces that are very difficult to manufacture by conventional machining operations. For example, propellers, airfoils, three-dimensional cans, punch presses, and dies might be manufactured using this stacking technique. An example of a layered mold of stacked sheet by DiMatteo is shown in Figure 1.9. In 1979, Nakagawa [17] used lamination techniques to produce various kinds of machine tools.

Photosculpture   The idea of photosculpture came about in the nineteenth century in an attempt to create exact three-dimensional replicas of objects, including human forms [18]. In 1860, Frenchman Francois Willeme success-fully designed a method to promote this technology. In this method, the object to be duplicated was placed in a circular room and simultaneously photo-graphed by 24 cameras placed equally about the circumference of the room. An artisan then used the silhouette in Willeme's studio to carve out 1/24th of a cylindrical portion of the figure. In an effort to simplify the labor-intensive carving process of Willeme's photosculpture, Baese [19] described a method using graduated light to expose photosensitive gelatin, which ex-pands in proportion to exposure when treated with water. Annular rings are then fixed to a support to make a duplicate of the object.

Morioka of Japan [20] developed a hybrid process combining the processes of topography and photosculpture. This method uses structured light to pho-tographically create contour lines of an object. The lines are then developed into sheets and cut and stacked or projected onto stock material for carving. Figure 1.10 shows the process. In 1951, Munoz proposed a system whose features are very similar to those of a stereolithography system. His system selectively exposes a transparent photoemulsion in a layerwise fashion, where each layer came from a cross section of the scanned object. These layers were

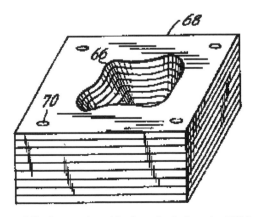

Figure 1.9 Layered mold of stacked sheet by DiMatteo.

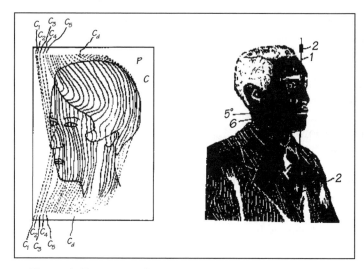

Figure 1.10   Process for manufacturing a relief by Morioka.

created by lowering a piston in a cylinder full of photoemulsion and fixing agent. After exposure and fixing, the resulting solid transparent cylinder contained an image of the object. Subsequently, this object could be further processed to create a 3D object.

1.3.2   Early Solid Freeform Fabrication

In 1968, Swainson [21] proposed a photochemical machining process to directly fabricate a plastic pattern by selective three-dimensional polymerization at the intersection of two laser beams. The pattern (object) is formed by either photochemically cross-linking or degrading a polymer by simultaneously exposing the polymer to the laser beams. While the process proved feasible theoretically, a viable practical process was never achieved.

In 1971, Ciraud [22] proposed a process that could be manufactured from a variety of materials that were particularly meltable. To produce an object, particles were applied to a matrix by gravity, magnetostatics, or electrostatics and then heated by a laser, electron beam, or plasma beam. The heating forced the particles to adhere to each other to form a continuous layer. Sometimes, more than one laser beam can be used to increase the strength of the bonding between the particles.

In 1981, Kodama [23] published an account of a functional photopolymer rapid prototyping system. In his process, a solid model was fabricated layer by layer, where each layer corresponded to a cross section of the model. Kodama studied the following three different methods for his process.

1. Using a mask to control exposure of the ultraviolet (UV) source and immersing the model downward into a liquid photopolymer vat in order to create a new layer
2. Using a mask as in method 1, but positioning the mask and exposure on the bottom of the vat and drawing the model upward to create a new layer
3. Immersing the model as in method 1, but using an *x-y* plotter and an optical fiber to expose the new layer

The three different methods are shown in Figure 1.11. In 1982, Herbert independently [24] proposed a process similar to Kodama. In this method, a UV laser beam was directed to a photopolymer layer by means of a mirror system on an *x-y* plotter (Fig. 1.12). A computer was used to command the laser beam across the photopolymer layer, the photopolymer layer was then lowered ($\sim$ 1 mm), and additional liquid photopolymer was then added to create a new layer.

Active Patents   The historical developments of rapid prototyping can also be traced through the list of patent developments. Table 1.1 shows a complete list of different active prototyping patents that cover existing commercial RP processes [25].

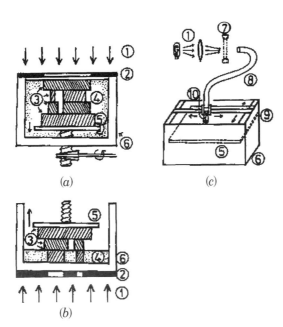

(a)

(c)

(b)

Figure 1.11   Early SFF concept, Codma's 3D photopolymer systems.

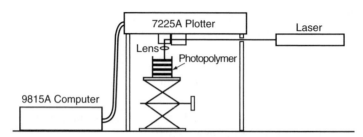

Figure 1.12 Early SFF concept, Herbert's photopolymer process.

1.3.3 Commercial Development

Early attempts to commercialize prototyping machines by Willeme, Swainson, DiMatteo, and others basically failed because of lack of experienced workers or funding.

Tables 1.2 and 1.3 show the chronology for commercial development of rapid prototyping processes in the United States, Europe, and Japan [25]. It is quite obvious from these tables that the United States leads the way in commercial development of rapid prototyping equipment. The United States also leads the way in developing different types of prototyping processes, compared to Japan and Europe. In Japan, with the exception of Kira, all other Japanese vendors use laser photopolymer techniques. 3D Systems is the only U.S. company that uses a photopolymer technique.

1.3.4 Chronology of RP Development

Table 1.4 shows an overall chronology of rapid mechanical prototyping. This chronology indicates some but not all of the major chronological events in the field [25].

1.4 DEVELOPMENT OF RP SYSTEMS

In this section, a series of new technologies that are available in the market is reviewed. These technologies allow "automatic" production of prototype parts directly from a solid model. The systems discussed here are commercial systems currently available in the United States.

1.4.1 3D Systems' Stereolithography Process

The first commercial rapid prototyping machine was introduced in the late 1980s by 3D Systems [26]. 3D Systems introduced the SLA 250, which was the first commercial implementation of the stereolithography technique de-

TABLE 1.1   Active RP Patents

Name	Title	Filed	Country
Householder	Molding process	December 1979	United States
Murutani	Optical molding method	May 1984	Japan
Masters	Computer-automated manufacturing process and system	July 1984	United States
Andre et al.	Apparatus for making a model of industrial part	July 1984	France
Hull	Apparatus for making three-dimensional objects by stereolithography	August 1984	United States
Promerantz et al.	Three-dimensional mapping and modeling apparatus	June 1986	Israel
Feygin	Apparatus and method for forming an integral object from laminations	June 1986	United States
Deckard	Method and apparatus for producing parts by selective sintering	October 1986	United States
Fudim	Method and apparatus for producing three-dimensional objects by photosolidification; radiating an uncured polymer	February 1987	United States
Arcella et al.	Casting shapes	March 1987	United States
Crump	Apparatus and method for creating three-dimensional objects	October 1989	United States
Helinski	Method and means for construction of three-dimensional articles by particle deposition	November 1989	United States
Marcus	Gas-phase selective beam deposition: three-dimensional, computer-controlled	December 1989	United States
Sachs et al.	Three-dimensional printing	December 1989	United States
Levet et al.	Method and apparatus for fabricating three-dimensional articles by thermal spray deposition	December 1989	United States
Penn	System, method, and process for making three-dimensional objects	June 1992	United States

TABLE 1.2   U.S. Commercial Development of RP Systems

Company	Process	Venture Start	Shipment	Notes
Aaroflex	Stereolithography	1995	N/A	License from DuPont
BPM	Ink jet	1989	1995	
DTM	Selective laser sintering	1987	1992	Operated a service bureau from 1990 to 1993
DuPont Somos	Stereolithography	1987	N/A	Lisensed to Teijin Seiki 1991, Aaroflex 1995
Helisys	Laminated object	1985	1991	Founded as Hydronetics
Light Sculpting	Photomasking	1986	N/A	Operates as a service bureau
Quadrax	Stereolithography	1990	1990	Technology acquired by 3D Systems in 1992
Sanders Prototyping	Ink jet	1994	1994	Partially developed at E-Systems
Soligen	3D printing	1991	1993	Operates as a service bureau
Stratasys	Fused deposition	1988	1991	
3D Systems	Stereolithography	1986	1988	First commercial shipment of equipment
Z Corp	MIT's 3DP	1994	1997	First commercial shipment

veloped by Charles Hull. The technique creates three-dimensional models from liquid photosensitive polymer resins that solidify when exposed to ultraviolet light. With reference to Figure 1.13, the model is built upon a platform situated just below the surface of a vat of photopolymer material. The photopolymer is a relatively low-viscosity liquid until it is exposed to UV light. The UV light causes the photopolymer molecules to polymerize, linking together to form a network. This process turns the liquid into a solid. The beam from an UV laser is directed into the vat by a mirror system that is controlled by a computer.

A software program slices the 3D CAD model into layers. The laser beam traces each layer onto the surface of the vat, which solidifies as a result of the exposure to UV light.

The solidified layer is then lowered into the vat so that another layer of liquid can be exposed to the laser. This process is repeated until all cross sections are built into a solid model of the original CAD model.

TABLE 1.3   **European and Japanese Commercial Development of RP Systems**

Company	Process	Venture Started	Shipment	Notes
CMET	Stereolithography	1988, Japan	1990	
Cubital	Photomasking	1987, Israel	1991	
Denken	Stereolithography	1985, Japan	1993	
DMEC	Stereolithography	1990, Japan	1990	
EOS	Stereolithography, Selective laser sintering	1989, Germany	1990	
Fockele & Schwarze	Stereolithography	1991, Germany	1994	Service bureau since 1992
KIRA	Laminated object	1992, Japan	1994	
Meiko	Stereolithography	1991, Japan	1994	
Mitsui	Stereolithography	1991, Japan	1991	
Sparx	Laminated object	Sweden	1994	Foam machine
Teijin Seiki	Stereolithography	1991, Japan	1992	License from DuPont
Ushio	Stereolithography	Japan	1994	

1.4.2   3D System's Selective Laser Sintering

The selective laser sintering (SLS) process [27], introduced by DTM, uses a laser to sinter a powdered material into the prototype shape. In many ways, SLS is similar to stereolithography apparatus (SLA) except that the laser is used to sinter and fuse powder rather than photocure a polymeric liquid. In this process, a thin layer of thermoplastic powder is spread by a roller over the surface of a build cylinder and heated to just below its melting point by infrared heating panels at the side of the cylinder. Then a laser sinters and fuses the desired pattern of the first slice of the object in the powder. Next, this first fused slice descends one object layer, the roller spreads out another layer of powder, and the process continues until the part is built. As the device builds layer after layer, the unsintered powder acts as support for the part under construction.

The key advantage of the SLS process is that it can make functional parts in essentially final materials. However, the system is more complex mechanically than stereolithography and other technologies. Surface finnish and part accuracy are not quite as good as those of other technologies, but material properties can be quite close to those of intrinsic materials. Since the prototyped parts are sintered, they are porous. Because of that, it may be necessary to infiltrate the part, especially metal parts, with another material to improve mechanical properties. Figure 1.14 shows a simple SLS diagram.

1.4.3   Helisys' Laminated Object Manufacturing

Helisys introduced the laminated object manufacturing (LOM) technique [28], which creates parts from laminated sheets of paper, plastic, or metal by a

TABLE 1.4   Chronology of RP Technology

Topography				Photosculpture
Perera patent filed	1937	1902	Baese patent filed	
Zang patent filed	1962	1922	Monteah patent filed	
Gaskin patent filed	1971	1933	Morioka patent filed	
Matsubara patent filed	1972	1940	Moriola patent filed	
Dimatteo patent filed	1974	1951	Munz patent filed	
Nakagawa laminated				

1981	Kodama publication
	Andre patent filed, Hull patent filed
1985	Helisys founded, Denken venture started
1986	Pomerantz patent filed, Peygin patent filed, Deckard patent filed, 3D Systems founded, Light Sculpting
1987	Fudim patent filed, Areella patent filed, Cubital, DTM founded, DuPont Somos Venture started
1988	First shipment by 3D Systems, CMET founded, Stratasys founded
1989	Crump patent filed, Melinski patent filed, Mareus patent filed, Sachs patent filed, EOS founded, BPM founded
1990	Levant patent filed, Quadrax founded DMEC founded
1991	Teijin Seiki venture started, Foeckele & Schwartze founded, Mitsui venture started
1992	Penn patent filed, Quadrax acquired by 3D Systems, KIRA venture started
1994	Sanders Prototype started, EOS commercialized EOSINT
1995	Aaroflex venture started, Japan's Unirapid machine
1996	BPM started selling its BPM technology, Kinergy of Singapore started its Zippy paper laminations system, 3D Systems sold its first Actua 2001; Schoff Development began to sell JPS System
1997	Aeromet was founded to develop LAM
1998	Beijing Yinhua Laser RP + Mold Making promoted its RP system, Optomec commercialized its LENS metal powder system
1999	3D Systems introduced ThermoJet system, Extrude Hone's ProMetal Division installed its first RTS-300, Roders began to sell its controlled metal buildup (CMB) machine
2000	Toyoda Machine Works of Japan started manufacturing and selling LOM-based system; Sanders Design International developed Rapid ToolMaker (RTM), Object Geometires of Israel announced Quadra, Precision Optical Manufacturing announced Direct Metal Deposition, Z-Corp introduced Z402 machine, Helysis went out of business, Cubic Technologies was formed to sell and service LOM
2001	CMET Inc. was formed out of the merger of, Teijn Seiki, NTT Data CMET, Aeroflex and Cubital ceased operation, 3D System acquired DTM and RPC Ltd.
2002	RSP Tooling received exclusive license for RSP technology
2003	Method and device for manufacturing three-dimensional bodies.

Selective inhibition of bonding of power particles for layered fabrication of 3-D objects.

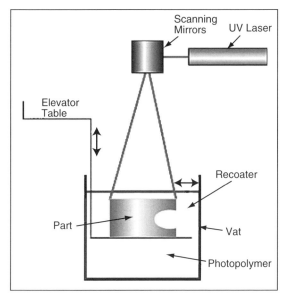

Figure 1.13 Stereolithography apparatus.

laser. This technique only requires a cut of the outline of each layer into a sheet of paper instead of cross hatching the entire part cross section. After each outline has been cut, the roll of paper is advanced, then a new layer is glued onto the stack, and the process is repeated until the part is made. After making the part, some trimming, hand finishing, and curing are needed. Companies requiring larger components, especially automobile industries, often prefer LOM over SLA or SLS processes. Figure 1.15 shows a LOM system.

1.4.4 Stratasys' Fused Deposition Modeling

Fused deposition modeling (FDM) [29] was introduced by Stratasys in 1991. FDM extrudes a thin stream of melted acrylonitrile–butadiene–styrene (ABS)

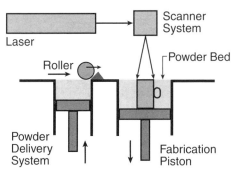

Figure 1.14 SLS diagram.

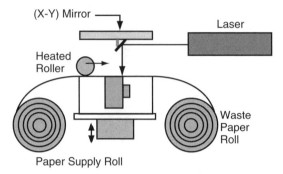

Figure 1.15   LOM system.

thermoplastic polymer or wax through an extruder head whose position is controlled by a computer. Parts are built up by moving the extruder head through the volume of the part. The process fabricates 3D parts from a buildup of 2D layers. Figure 1.16 shows the extrusion head of an FDM machine.

1.4.5   Solidscape's 3D Printing

Solidscape (formerly known as Sanders Prototype, Inc. (SPI) [30] introduced 3D printing technology that uses dual ink printer heads to deposit both thermoplastic part material and a wax support structure. The device uses a milling operation after each layer is deposited to create very thin, precise layers. We shall study this process in more detail in Chapter 3. The principles and operation of Solidscape's RP ModelMaker will also be described in the case

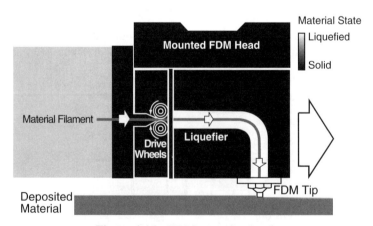

Figure 1.16   FDM extrusion head.

study section of this chapter (Section 1.9). Figure 1.17 shows the 3D printing process.

1.5   APPLICATIONS IN EDUCATION AND INDUSTRY

Since the delivery of the first commercial machine in 1988, RP has grown into an integral part of the new product development process. The use of RP has reduced time to market, cut trial costs, and improved product quality by giving design and manufacturing professionals the tool to verify and fine tune designs before committing them to expensive tooling and fabrication.

1.5.1   Application in Product Development

The world has already entered into a new era of global competition for providing products and services. Rapid acceleration of new and emerging technologies is fueling this growth in all aspects of business. Companies engaged in product development and manufacturing are in tremendous competition to bring a product to market faster, cheaper, and with both higher quality and functionality. Reducing the timeline for product development saves money in the overall time-to-market scenario. RP is the technology that helps companies reduce the cycle of product development and also facilitate making design improvements earlier in the process where changes are less expensive.

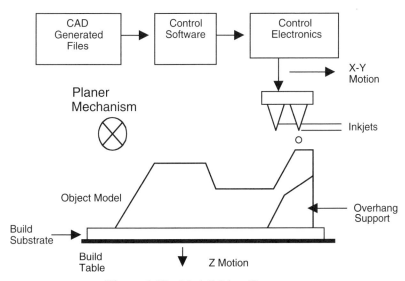

Figure 1.17   ModelMaker II process.

The impact RP can have on product development [31] is shown in Figure 1.18.

Figure 1.18 charts a design process for product development. The design process includes the following steps:

1. Design concepts
2. Parametric design
3. Analysis and optimization
4. Creation of prototype
5. Prototype testing and evaluation
6. Comparing to design criteria
7. Final product or prototype

The first step in design arises out of the need for a product to meet customer expectations. It requires the engineer to use experience, knowledge, and ingenuity to work out a preliminary concept of what the design should look like.

The second step involves parametric design of the concept. CAD techniques using software packages such as AutoCAD, Pro/Engineer, or any number of commercially available CAD programs are used in this step.

Once a preliminary design is accomplished the design is analyzed and optimized. The optimized design is then prototyped, tested, and a final preproduction part is fabricated. Design and manufacturing companies use RP to produce models and prototype injection molded components and castings that are used in everything from telephones to printers to computers to automobiles to airplane parts as well as for instrument panels and medical applications. Companies are using RP for a variety of reasons such as improving time to market, creating visualization tools, and reducing waste in the design process by identifying flaws in the design early in the manufacturing cycle.

Although the applications of RP to product development are widespread, most applications can be subdivided into three major categories: prototype

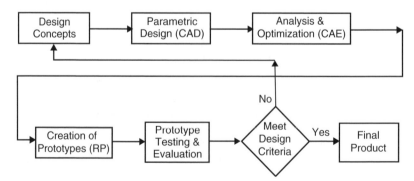

Figure 1.18 Impact of RP on product development.

design evaluation, prototype function verification, and prototypes for manufacturing process validation [32].

Prototype Design Evaluation Most modern designs use solid modeling techniques to facilitate the design evaluation by providing such functions as viewing, shading, rotating, and scaling. However, there is no substitute for a design evaluation where the design can be held and felt. Without the actual prototype, no design can be accurately visualized. The visualization of prototypes can help reduce design error and help reduce analysis of the product.

Prototype for Function Verification Once a prototype is made, its functionality (such as checking the practicality of assembly), kinematic, and aerodynamic performance need to be verified. Since most modern parts are complex, and most require assembly, it is important to verify that a product can be assembled from its components and disassembled easily. Prototype components and assemblies made from various RP machines can save time and cost for function verification.

Prototypes for Manufacturing Process Validation Verification of kinematic performance is necessary for moving parts in an assembly. Part motion is hindered by interference of moving parts in an assembly. Prototype parts can be tested for checking interference during assembly, thus saving time and cost. Prototypes for casting or machining can be properly evaluated for process compatibility only by having a physical model. RP makes this possible at relatively low cost.

Aerodynamic performance of products is best evaluated using a wind tunnel. The geometric shapes of prototypes influence their aerodynamic performance. However, not all prototypes have the strength necessary for wind tunnel testing. Prototypes made from ABS plastic from Stratasys have done very well in wind tunnel testing. In order to verify characteristics such as strength, operational temperature limits, fatigue, and corrosion resistance, prototypes should be made from the same type of materials as the finished product.

1.5.2 Application in Reverse Engineering

Reverse engineering (RE) is the science of taking an existing physical model and reproducing its surface geometry in a 3D data file on a CAD system [33]. In many cases, only the physical model of an object is available. Examples of such situations include hand-made prototypes, craft works, reproduction of old engineering objects, and sculptured bodies found in medical and dental applications. In order to facilitate CAM of these physical models, it is essential to establish a CAD model [34, 35]. RE is the quickest way to get 3D data into any computer system. It is similar to having a low-cost but accurate X-ray machine for parts or having a 3D copier.

The example of the application of RP in reverse engineering will be shown by the following example problem.

Example 1.1   Reverse Engineering a Christmas Tree Using Coordinate Measurement Machine

Problem Objective

The objective of this problem was to study the building of a "Christmas tree" calibration device using a coordinate measurement machine (CMM). Recently, reverse engineering technology has emerged as a valuable tool in making prototypes of parts for which there is no design data available. In addition to the build parameters, the amount of point data generated, surface finish, and accuracy of the regenerated surfaces are also investigated.

Description of Work Done on Project

The project was done using an experimental system. The experimental system consisted of an SLA-250 stereolithography machine, a CMM (Microval-343), personal computer, and a printer. The test part was a Christmas tree that was created using an SLA-250 stereolithography machine. Although design data for the part was available, we wanted to see whether we could recreate the part using the principles of reverse engineering.

The test part was put under the CMM and, after initial configuration of the CMM and the PC- DMIS software for Windows, measurements of the part were made manually. The PC-DMIS software has a unique database structure that continuously manages the measurement environment. We were able to move effortlessly among *learn, edit,* and *execution* modes. Measurements of the height, width, and thickness of the part were taken using the touch probe of the CMM. The data analysis was made using the PC-DMIS software.

The point clouds created from the measurement data were used to create a part file. The part file was then converted to an IGES (initial graphics exchange specification) file, which is a neutral file format for CAD programs. From the IGES file, we were able to re-create the original part and prototype it. We can even make the part in metals using investment casting technique. We were surprised at the accuracy of the measured data, as compared to the original design data. So as long as the surface is clean, the accuracy of the measurement is very good. While laser scanning technology improves the speed of measurement, there is no doubt that the accuracy of manual measurement is easier from the standpoint of data reduction. Figure 1.19 shows a reverse-engineered Christmas tree. The details of reverse engineering will be discussed in Chapter 7.

1.5.3   Application in Casting and Pattern Making

Another important application of RP is in casting and pattern making. In this case, the pattern and the cores for the casting process are made using the RP process. Investment casting, for example, is a very convenient process employed to make such things as jewelry, dental fixtures, small mechanical components, and blades for turbine engines. Investment casting offers an

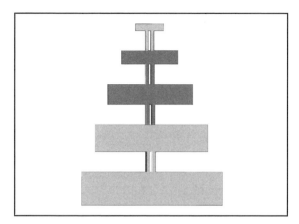

Figure 1.19   Reverse-engineered Christmas tree.

alternative to machining, especially in cases where the shapes and details are extremely irregular. Investment casting allows fabricators to make complex (very detailed) parts quite simply. In investment casting, a model made of wax is dipped in a ceramic slurry. This creates a ceramic shell around the wax model (pattern). The wax is then melted out of the shell, the shell is hardened by heating, and the molten metal of choice is then poured in. Metals used for this process include 14K and 18K gold, bronze, and silver to list a few. For this particular case study, a bronze model was made using investment casting [36]. Figure 1.20 shows the model. This case study will be explored in more detail in a later chapter.

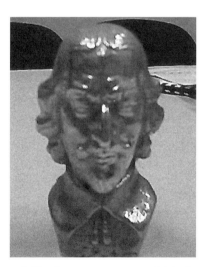

Figure 1.20   Investment cast model in bronze.

Biomet, a maker of surgical implants, has used FDM rapid prototyping to produce 60 implantable casting designs per month. The company has also produced another 600 to 1000 castings monthly, using injection-wax patterns from steel tools, due to the volume requirements needed from a single design.

Engineers have tested many designs using ABS patterns, and they have been able to achieve accurate castings in as little as 2 weeks.

1.5.4 Application in Rapid Tooling

Rapid tooling (RT) is a new term that has not yet been clearly defined. It was originally linked to rapid prototyping but has since been used to define techniques and processes that lead to making tools quickly. These may include direct tooling, single-reverse tooling, double-reverse tooling, and triple-reverse tooling. With the increase of reversals, the durability of a product can be improved, but the product cost increases while its precision decreases.

The need for faster and less expensive tooling solution has resulted in more than 20 RT methods. Many more RT methods, processes, and systems are being developed, some of which are yet to be well understood in industry.

As an example, Toyota Motor Company has used FDM rapid prototyping to cut $200,000 in tooling costs by eliminating the prototype tooling for the right-hand side mirror housing. Toyota also saved $300,000 in tooling costs for the four doorhandles, which were created from FDM masters, rather than by CNC machining. It is reported that FDM rapid prototyping has saved Toyota $2 million on Toyota model cars in recent years [9].

1.5.5 Application in Medicine

Medical applications of rapid prototyping are increasing everyday, making the future of RP more and more promising. Possibly the most obvious application is as a means to design, develop, and manufacture medical devices and instrumentation. This is an outgrowth of other recognized engineering applications of the technology. Any field where it is imperative to decrease product development time is an excellent prospect for rapid prototyping. It therefore follows that since human lives depend on the quality and ease of use of numerous medical products, there is extra incentive to use rapid prototyping in their development. Examples of medical instruments designed using the technology include retractors, scalpels, surgical fasteners, display systems, and many other devices [1]. However, the use of RP models extends beyond instrumentation to include surgical implants and anatomical models. The human brain is reconstructed from magnetic resonance imaging (MRI) data in Figure 1.21. Velocity^2 Pro segmentation features were used to isolate the cortex, pallidum, cerebellum, and brainstem. The separate structures were then rendered in a CAD program so the file could be easily constructed using rapid prototyping [37]. This type of procedure can be used to prepare for delicate surgery or to determine other treatment options.

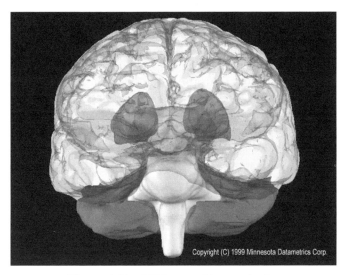

Figure 1.21   MRI scan of human brain.

1.5.6   Application in Rapid Manufacturing

The methods of rapid prototyping and rapid tooling are having profound effects on companies doing design and manufacturing. Some companies are extending the envelop of RP and RT into something called rapid manufacturing (RM). Using forward thinking, companies have successfully used RP and RT for the production of finished manufactured parts. Many believe that this practice of making final parts, which is rapid manufacturing, is the future of RP.

There are still a lot of challenges to RM such as the limitations of speed, materials, and accuracy of RP machines. However, as companies demonstrate more successful applications, more and more efforts will be put into the rapid manufacturing technology, and many of the obstacles to RM will be overcome.

The aerospace industry is one of the biggest users of RP. However, the industry imposes stringent quality controls on all the prototypes used for various applications. Yet, Boeing's Rocketdyne Division has successfully used RP and RT to manufacture hundreds of parts for applications on the International Space Station and space shuttle fleet [9]. RP manufactured parts have also been used for F-18 fighter jets. Rocketdyne has manufactured these parts using 3D Systems' SLS process.

The students of the SAE (Society of Automative Engineers) Formula 1 team at Loyola Marymount University (LMU) routinely use rapid prototyped parts for their vehicle, and some are produced as final production parts for their cars built to compete in races. Some of the parts include intake manifold, side-view mirrors, and various molds and models for the vehicle.

Another application of rapid manufacturing is in the Invisalign technology. Invisalign is the invisible new way to straighten teeth without braces. It uses a series of clear, removable aligners to gradually straighten teeth, without metal or wires. Invisalign uses 3-D computer imaging technology to depict the complete treatment plan from the initial position to the final desired position from which a series of custom-made Aligners are produced. Each Aligner moves teeth incrementally and is worn for about 2 weeks, then replaced by the next in the series until the final position is achieved. These Aligners are made by RP machines (mostly by SLA machines of 3D Systems).

1.6   CASE STUDY: FABRICATING A PROTOTYPE USING 3D PRINTING

ModelMaker from Solidscape produces parts by printing successive layers of wax-type build and support materials on foam substrate attached to an elevator table. Each layer consists of a horizontal cross section of the model. After a layer is completed, the table is lowered slightly, and the next layer is printed on top of the last one. In this manner the thin cross sections are stacked on top of each other to build up the three-dimensional model. Figure 1.22 shows the functional block diagram of ModelMaker [30].

Many models have an overhang at some point, that is, a layer that is larger than the one below it. This can create a problem since there is nothing to

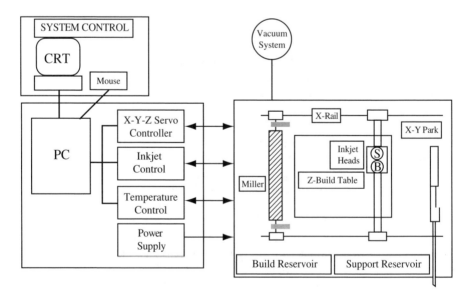

Figure 1.22   Functional block diagram.

support the overhung portion when it is printed; instead, the overhung portion will fall down to the table surface below. To prevent this, temporary support material must be added to the lower layers wherever an overhang will occur. The support material is removed after the model is completed.

Before a model can actually be built, a binary data (BIN) file must be prepared. This file contains the data that describes the physical geometry of the part. It defines each drawn vector of each successive layer of the model and includes both build and support materials.

ModelMaker also requires a build process parameter (BLD) file. This file contains the data that describes how the part will be built. A default file with the most commonly used settings is provided. However, a customized .BLD file can be created to meet particular requirements.

A software program called Model Works creates .BIN files from CAD data. This data can be a three-dimensional STL file or a series of two-dimensional Hewlett-Packard graphics language (HPGL) files (where each file represents a different cross section of the part.) The Model Works program can check the data for errors, allow simple editing, orient the part in various ways, generate support structures, and so forth. Other third-party programs can also generate .BIN files for the ModelMaker. Some of these programs can also create a customized .BLD file to match each .BIN files.

Once the .BIN file is ready, ModelMaker can build the part. A program called MM reads the .BIN file and controls the machine to print the successive layers. Separate jets are used to print the build and support materials. A jet check station allows the jets to be tested for proper operation and to be cleaned if necessary, all automatically. As each layer is completed, a miller is used to trim the surface of the model to ensure a flat uniform surface for the next layer.

The user can monitor the process through the computer display and a log file that is automatically kept on disk. Modifications to the build process can be made through the computer keyboard and a process parameter file.

The support material is used to support overhangs and cavities in the model during the model build sequence. The build and support materials are deposited onto the build substrate as a series of uniformly spaced "microdroplets." These droplets may be placed at any desired location upon the build substrate. The droplets adhere to each other during a liquid-to-solid phase transition to form a uniform mass. The drying process is fast enough to allow the milling of the layers immediately following the deposition cycle.

PROCEDURE   The three-dimensional solid model of the part was produced using Pro/Engineer. The model is saved as a binary .STL file and sliced into layers for the prototyping process. The file is then sent to the RP machine. The machine is allowed to proceed with the model making process until the part is complete. The finished model contains support material that is ultimately removed. This is done by placing the model in a solution to dissolve the support material and the build substrate.

Of the several prototypes that were made using the Sanders prototype (ModelMaker), only two will be presented here. Figure 1.23 shows the first part. The part was created in a Sun Microsystems's Sparcstation 5 computer using Pro/Engineer solid modeling software. The .STL file was then transferred to ModelMaker where it was prototyped. Figure 1.24 shows a complex mechanical part. Because the part came from a defense firm, we cannot release the name of the part. The form, fit, and function of both parts appear excellent. Using a CMM, the dimensional accuracy of parts produced using Model Maker were measured and analyzed. The measurement and analysis were done at NASA's Jet Propulsion Laboratory (JPL).

The prototypes made by the Sanders prototype were compared to those made by 3D Systems' stereolithography process. Except for the material strength, the prototypes from the Sanders prototype compared favorably with those from 3D Systems insofar as the form, fit, and finish were concerned. Generally speaking, for small parts, the Sanders prototype system can be set to create more accurate prototypes and patterns than the SLA machines from 3D Systems.

1.7   SUMMARY

In this chapter, the basic concepts of rapid prototyping were introduced and discussed. Rapid prototyping refers to fabrication of a physical, three-dimensional model using a special class of machine technology. Some of the advantages of RP are reduced lead time to bring a product to the market, improved part geometry, and early detection of any design error. Many industries are using RP in their environment for producing quality products.

Figure 1.23   Sanders prototype (ModelMaker), Part 1.

Figure 1.24   Sanders prototype (ModelMaker), Part 2.

The basic RP process involves creating a CAD model of a part, converting the drawing file to STL format, slicing the file into two-dimensional cross sections, growing the part, and then cleaning and finishing the part. Although the modern concept of RP was conceived and promoted in the 1980s, the origin of this technology dates back to the 1890s. The two earlier areas of *topography* and *photosculpture* also played a part in the development of the modern RP concept.

Rapid prototyping for electromechanical part design came about in 1987 with the stereolithography process, developed by 3D Systems. The stereolithography process solidifies layers of liquid photopolymers when exposed to ultraviolet rays. 3D Systems then commercialized its SLA (stereolithography apparatus) machine worldwide. Japan's NTT Data CMET and Sony/D-MEC commercialized their versions of the stereolithography machines in 1988 and 1989, respectively. Then 1991 saw the emergence of fused deposition modeling from Stratasys, solid ground curing from Cubital, and laminated object manufacturing from Helysis. Both Cubital and Helisys are out of business now, but many of their machines are still in use. Since then, many more systems have been introduced including selective laser sintering, direct shell production casting, and 3D printing.

The early applications of rapid prototyping were geared toward product development. Most applications can be subdivided into three main categories: prototype design evaluation, prototype for function verification, and prototype for manufacturing process validation. In addition, rapid prototyping is being used in reverse engineering, casting and pattern making, rapid tooling, and in medical applications.

Rapid prototyping is a new technology that is impacting the way products are made. Despite some challenges and difficulties, rapid prototyping processes and activities will grow in the future. The process will continue to move from rapid prototyping to rapid manufacturing in the near future as more applications become apparent and faster machines become available.

PROBLEMS

1.1. What is a prototype? Why do you need a prototype? What are potential uses for a prototype?

1.2. What are the questions that need to be asked when considering physical prototypes?

1.3. What are the general classes of prototypes?

1.4. Define rapid prototyping (RP).

1.5. What are the major advantages of RP?

1.6. What is the difference between RP and conventional machining?

1.7. What are the main steps in making a prototype using RP?

1.8. Name five RP machine producers and describe the principles of each system.

1.9. Show the positive impact of RP on the product development cycle using a few diagrams.

REFERENCES

1. M. Burns, *Automated Fabrication: Improving Productivity in Manufacturing,* Prentice-Hall, Englewood Cliffs, NJ, 1993.
2. S. Ashley, *Rapid Prototyping Systems—Special Report,* Mechanical Engineering, 1991, pp. 34–43.
3. S. Ashley, *Rapid Prototyping is Coming of Age.* Mechanical Engineering, July, 1995, pp. 62–68.
4. P. Jacobs, *Rapid Prototyping and Manufacturing & Fundamentals of Stereolithography.* SME Publications, 1992.
5. C. L. Thomas, *An Introduction to Rapid Prototyping.* SDC Publications, Mission, Kansas, USA, 1996.
6. M. P. Groover, *Fundamentals of Modern Manufacturing: Materials Processes and Systems,* Prentice-Hall, Englewood Cliffs, NJ, 1996.
7. Ch. Kai and L. Fai, *Rapid Prototyping Principles & Applications in Manufacturing,* Wiley, New York, 1997.
8. P. Jacobs, *Rapid Prototyping and Manufacturing & Fundamentals of Stereolithography.* SME Publications, Dearborn, Michigan, 1992.
9. T. Wohlers, Rapid Prototyping, Tooling and Manufacturing State of the Industry, *Wholers Report 2004,* Wohlers Associates, Fort Collins, CO, 2004.
10. J. Beaman, H. Marcus, D. Bourcll, J. Barlow, R. Crawford, and K. McAlea, *Solid Freedom Fabrication: A New Direction in Manufacturing,* Kluuser Academic, Boston, 1996.
11. J. E. Blanther, U.S. Patent 473,901, 1892.
12. B. V. Perera, U.S. Patent 2,189,592, 1940.

13. E. E. Zang, U.S. Patent 3,137,080, 1964.

14. T. A. Gaskin, U.S. Patent 3,751,827, 1973.

15. K. Matsubara, Japanese Kokai Patent Application, Sho 51 [1976]-10813, 1974.

16. P. L. DiMatteo, U.S. Patent 3,932,923, 1976.

17. T. Nakagawa, Blanking Tool by Stacked Bainite Steel Plates, *Press Technique:* 93-101, 1979.

18. M. Bogart, In Art the End Don't Always Justify the Means, *Smithsonian,* 104–110, 1979.

19. C. Baese, U.S. Patent 774,549, 1904.

20. I. Morioka, U.S. Patent 2,350,796, 1944.

21. W. K. Swainson, U.S. Patent 4,041,476, 1977.

22. P. A. Ciraud, *FRG Disclosure Publication 2263777,* 1972.

23. H. Kodama, Automatic Method for Fabricating a Three Dimensional Plastic Model with Photo-Hardening Polymer, *Review of Scientific Instrumentation,* 1770–1773, 1981.

24. A. J. Herbert, Solid Object Generation, *Journal of Applied Photographic Engineering,* 8[4], 1882.

25. World Technology Evaluation Center (WTEC), Baltimore, MD, 2002.

26. 3D Systems, *3D Systems Products Brochure,* Valencia, CA, 1996.

27. K. Otto and K. Wood, Product Design: Techniques in Reverse Engineering and New Product Development, Prentice-Hall, Englewood Cliffs, NJ, 2001.

28. Helisys, *Selective Laser Sintering Products Brochure,* Helisys, Austin, TX, 1998.

29. Stratasys, *LOM Products Brochure,* Helysis, Carson, CA, 1998.

30. Stratasys, *Stratasys Products Brochure,* Eden Prairie, Stratasys, 1998.

31. *Solidscape Products Brochure,* Merrimack, NH, 1997.

32. J. Lee, *Principles of CAD/CAM/CAE Systems,* Addison Wesley, Reading, MA, 1999.

33. J. Singh, *Rapid Reverse Engineering to Rapid Prototyping: A Case Study, Proceedings on Reverse Engineering,* SME, Newport Beach, CA, 1998.

34. C. Schoene and J. Hoffmann, Reverse Engineering Based on Multi Axis Digitized Data, Proceedings of the International Conference on Manufacturing Automation, Hong Kong, 1997.

35. G. C. Lin and L. C. Chen, A Vision-Aided Reverse Engineering Approach to Reconstruct Free-Form Surfaces. Proceedings of CAD/CAM, Robotics and Factories of the Future, London, 1996.

36. R. Noorani, S. Dorman, and K. Grote, Global Product Development Using Reverse Engineering, Rapid Prototyping and Investment Casting, Proceedings of the International Conference on Manufacturing, Dhaka, Bangladesh, 2000.

2

RAPID PROTOTYPING PROCESS

The objective of rapid prototyping is to quickly fabricate any complex-shaped, three-dimensional part from CAD data. Rapid prototyping is an example of an additive fabrication process. In this method, a solid CAD model is electronically sectioned into layers of predetermined thickness. These sections define the shape of the part collectively.

The focus of this chapter is on those system elements that affect the shape of the part: the CAD file, the STL (stereolithography) file, problems and repairs of STL files, and other file formats. In short, the modeling principles of rapid prototyping will be discussed in this chapter.

2.1 PRINCIPLES OF AUTOMATED PROCESSES

Rapid prototyping is essentially a part of *automated fabrication,* a technology that lets us make three-dimensional parts from digital designs [1]. There are several advantages of *automated fabrication* over manual fabrication and molding processes. Some of these advantages are computer-aided design, quick design changes, and precise dimensioning. Fabrication processes, manual or automated, can be classified as subtractive, additive, or formative. These processes are shown in Figure 2.1.

Subtractive Process In this process, one starts with a solid block of material larger than the final size of the finished object, and then material is removed slowly until the desired shape is reached.

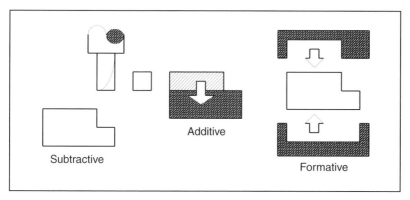

Figure 2.1 Three types of fundamental fabrication processes.

Subtractive processes include most forms of machining processes—computer numerical control (CNC) or otherwise. Most widely used examples include milling, turning, drilling, planning, sawing, grinding, electrical discharge machining (EDM), laser cutting, water-jet cutting, and many other methods.

Additive Process   Unlike the subtractive process, this process involves manipulation of material so that successive pieces of it combine in the right form to produce the desired object.

The rapid prototyping process (layered manufacturing) falls into the additive fabrication category. Figure 2.2 shows the additive layer-by-layer process of rapid prototyping. Examples of RP processes include SLA, FDM, LOM, SLS, solid ground curing (SGC), direct shell production casting (DSPC), and 3D printing (3DP).

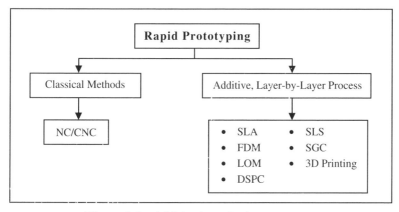

Figure 2.2 Additive layer-by-layer process.

Formative Process   In this process, mechanical forces are applied to material so as to form the desired shape.

Examples of the formative fabrication process include bending, forging, electromagnetic forming, and plastic injection molding.

Two or three of these processes can be combined to form a *hybrid* process. Hybrid processes are expected to contribute significantly to the production of goods in the future. Progressive press working is an example of hybrid machines that combine two or more fabrication processes. In progressive press working a hybrid of subtractive (as in blanking or punching) and formative (as in bending and forming) is used.

2.2   RP FUNDAMENTALS

All prototypes made with both the current and evolving RP processes have several features in common [2, 3]. A solid or surface CAD model is electronically sectioned into layers of predetermined thickness. These sections define the shape of the part collectively. Information about each section is then electronically transmitted to the RP machine layer by layer. The RP machine processes materials only at "solid" areas of the section. Subsequent layers are sequentially processed until the part is complete. It is this sequential, layered, or lithographic approach to parts manufacturing that defines RP. The RP process basically uses the following steps to make prototypes:

1. Create a CAD model of the design.
2. Convert the CAD model to STL file format.
3. Slice the STL file into 2D cross-sectional layers.
4. Grow the prototype.
5. Postprocessing.

The five-step process is shown in Figure 2.3.

2.2.1   Creation of Solid Models

The first step in creating a prototype is the creation of a CAD solid model. RP requires that we make a fully closed, water-tight model such that even if we were to pour water into the volume of the model, it would not leak.

A solid is a volume completely bounded by surfaces, which means the edges of all surfaces must be coincident with one, and only one, other surface edge. Unlike wire-frame and surface modeling, solid modeling stores volume information. A CAD solid model not only captures the complete geometry of an object, it can also differentiate the inside and the outside of the space of that object. Many other volume-related data can be obtained from the model.

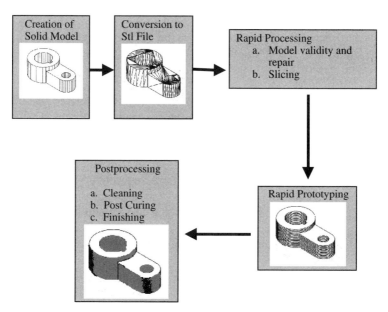

Figure 2.3   Five-step process of rapid prototyping.

Solid models can be created using a CAD software package such as AutoCAD, Pro/Engineer, CATIA, Solid Works, or many other commercially available solid modeling programs.

2.2.2   Conversion to STL File

Once a solid model is created and saved, it is then converted to a special file format known as STL (stereolithography) [4]. This file format originated from 3D Systems, which pioneered the stereolithography process. Actually, the Albert Consulting Group under contract to 3D Systems developed the STL file format to support the new revolutionary manufacturing technology, called stereolithography. Though not ideal, it is sufficient to meet the needs of today's rapid prototyping technology, which generally build monomaterial parts. The success of this file format has been impressive. Today, a decade later, the STL file format remains the de facto standard for the rapid prototyping industry.

The success of the STL file format is due to its sufficiency, its simplicity, and its monopoly [5]. Its mathematical sufficiency stems from the fact that it describes a solid object using a boundary representation (B-rep) technique. An STL file format represents the virtual CAD model of the object to be prototyped as a collection of triangular facets. These triangular facets, when taken together, describe a polyhedral approximation of the objects' surface, that is, a polyhedral approximation of the boundary between material and nonmaterial. In short, an STL file is nothing more than a list of x, y, and z coordinate triplets that describe a connected set of triangular facets. It also includes the direction of the normal vector for each triangle, which points to the outer surface of the model.

Most CAD/CAM software vendors supply the STL file interface. Since 1990, many CAD/CAM vendors have developed and integrated this interface into their systems. Tessellation is the process of approximating a surface by triangular facets. The CAD STL file interface performs surface tessellation and then outputs the facet information to either a binary or ASCII STL file format.

The output of STL file formats can be expressed in binary or ASCII format. The characteristics of binary and ASCII STL outputs are shown in Table 2.1.

Figures 2.4 and 2.5 show a binary STL file format and an ASCII STL file format, respectively.

There are three steps to STL file creation [6]:

1. Select the part(s) to be converted to STL representation.
2. Set the various tolerance parameters for the process.
3. Create a triangular representation of the geometry into an output file.

As mentioned, both surface and solid models can be converted to STL file formats. However, it is very difficult to create STL files from surface models. When processing surface models, the following steps should be followed:

1. Determine all surface adjacencies.
2. Triangulate each surface.
3. Make sure all edge vertices match.
4. Determine a normal that points to the outside of the model for each surface.
5. Output triangles and *normals* to an output file.

During creation of the STL file, the following interface options must be addressed:

TABLE 2.1   Output Types: Binary Versus ASCII STL

Binary	ASCII
Default output type	Referred to as human-readable format
Referred to as machine-readable format	Easily read and understood by humans
More compact and efficient, easier to move through network/transmit	Not very efficient, slower to process, larger file sizes
Not easily read or understood by humans without some translation	Not recommended if moving files through a network

```
Address        Length       Type              Description
   0             80         char        Header information
   80             4         long        Number of facets in solid

First facet (50 bytes):
   84             4         float       Normal (x component)
   88             4         float       Normal (y component)
   92             4         float       Normal (z component)
   96             4         float       Vertex 1 (x component)
  100             4         float       Vertex 1 (y component)
  104             4         float       Vertex 1 (z component)
  108             4         float       Vertex 2 (x component)
  112             4         float       Vertex 2 (y component)
  116             4         float       Vertex 2 (z component)
  120             4         float       Vertex 3 (x component)
  124             4         float       Vertex 3 (y component)
  128             4         float       Vertex 3 (z component)
  132             2         short       Attribute info. (not used)

Second facet (50 bytes):
134...
```

Figure 2.4 Binary STL format.

1. Triangulation tolerance
2. Adjacency tolerance
3. Auto-normal generation (on/off)
4. Normal display (on/off)
5. Triangle display (on/off)
6. Header information (text)

```
    SOLID Test_Part_567.Mainfold_z
    FACET NORMAL   0.000000e+00 1.000000e+00 2.634149e−09
OUTER LOOP
    VERTEX         3.000000e+00 1.400000e+00 4.000000e+00
    VERTEX         4.000000e+00 1.400000e+00 4.000000e+00
    VERTEX         4.000000e+00 1.400000e+00 3.000000e+00
    ENDLOOP
    ENDFACET
    FACET NORMAL...
    :
    <remaining facets>
    :
    ENDSOLID
```

Figure 2.5 ASCII STL format.

Some of the features of triangulation tolerance are as follows:

1. Determines how smooth the approximation of the surface or solid will be. In other words, how close the triangles approximate the surface.
   - How close the sides of the triangles that lie along the edges are to the actual edges of the surface.
   - Usually set to one half the desired accuracy of the RP process being utilized.
   - Default is set at 0.0025 in., or 0.05 mm.

The dramatic effects of decreasing the values of triangle tolerances are shown in Figure 2.6.

Some of the features of adjacency tolerance are as follows:

1. It does not affect processing of solids.
2. The default value is 0.005 in., or 0.12 mm.

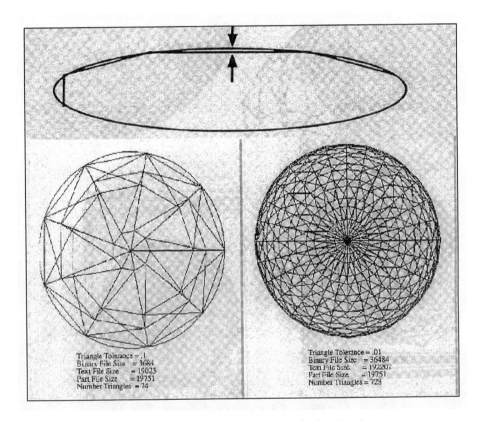

Figure 2.6   Effects of decreasing values of triangle tolerances.

3. The system uses this value to determine if two surfaces will be attached to one another.
4. Edges whose length is smaller than the adjacency tolerance can cause adjacency problems.

The effects of adjacency tolerance are shown in Figure 2.7.
Some of the features of auto-normal generation are as follows:

1. It does not affect processing of solids.
2. Choose a base surface, check the normal, and calculate all others from this surface.
3. Default should be on.

Example 2.1
For the object shown in Figure 2.8:

(a) Draw the part using Pro/Engineer or a similar CAD/CAM system.
(b) Create STL ASCII file with chord height of 1.0 and 0.1.
(c) Discuss the changes that take place when chord height is varied (file size, number of triangles, etc.).

Solution

(a) To generate the solid object, a simple rectangular *protrusion* was first created. Using the *cut* command in the file menu, the rectangular sections were *blind* cut to the specified depths. The circles were then cut completely through the part. The solid object is shown in Figure 2.9.
(b) The part was saved in an ASCII STL file format with the specified chord heights and is shown in Figure 2.10.
(c) Varying the chord height changes the number of triangles created. When the chord height is decreased, there are more triangles, thus a larger file size. The smaller the chord height, the more accurate the

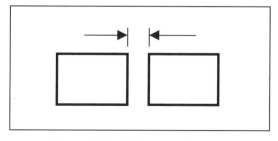

Figure 2.7 Adjacency tolerance.

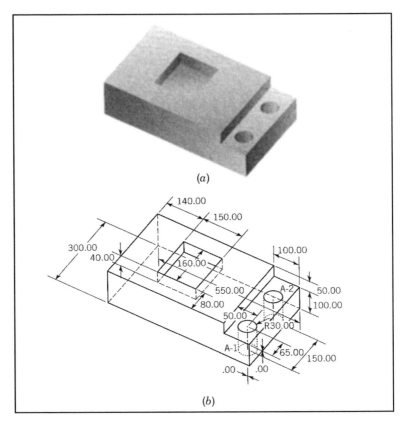

Figure 2.8   Part geometry.

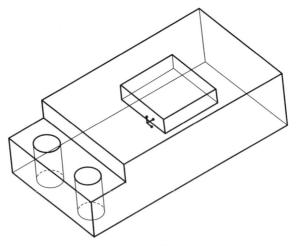

Figure 2.9   Solid object.

STL ASCII File Format

```
solid PROBLEM1
 facet normal 0.000000e+00 −0.000000e+00 −1.000000e+00
  outer loop
    vertex 5.500000e+02 1.00000e+02 −1.500000e+02
    vertex 5.500000e+02 1.00000e+00 −1.500000e+02
    vertex 0.000000e+00 0.00000e+00 −1.500000e+02
  endloop
 endfacet
 facet normal 0.000000e+00 −1.000000e+00 0.000000e+00
  outer loop
    vertex 0.000000e+00 0.00000e+00 1.500000e+02
    vertex 0.000000e+00 0.00000e+00 −1.500000e+02
    vertex 5.500000e+02 0.00000e+00 −1.500000e+02
  endloop
 endfacet
 facet normal 0.000000e+00 −0.000000e+00 −1.000000e+00
  outer loop
    vertex 4.500000e+02 1.00000e+02 −1.500000e+02
    vertex 5.500000e+02 1.00000e+02 −1.500000e+02
    vertex 0.000000e+00 0.00000e+00 −1.500000e+02
  endloop
 endfacet
 facet normal 0.000000e+00 −0.000000e+00 −1.000000e+00
  outer loop
    vertex 4.500000e+02 1.50000e+02 −1.500000e+02
    vertex 4.500000e+02 1.00000e+02 −1.500000e+02
    vertex 0.000000e+00 0.00000e+00 −1.500000e+02
  endloop
 endfacet
```

Figure 2.10   STL ASCII.

object. However, the more accurate model will take longer to produce. Figure 2.11 shows the object sliced at different chord heights.

2.2.3   Slicing the File

The file is taken from its 3D model surfaces and converted to many triangles, a step referred to as *slicing*. The more complex the object, the more triangles are required, and thus the bigger the file that makes up the CAD model as well as a support structure for the part to be grown on. The sliced object is saved as an STL file and is now in a format the RP computer recognizes.

2.2.4   Making or "Growing" the Prototype

The part is submitted to the RP computer and the machine runs until the part is complete. RP machines build one layer at a time from polymers, paper, or

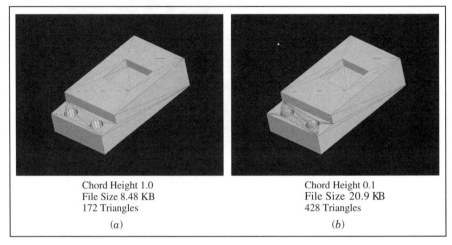

Chord Height 1.0	Chord Height 0.1
File Size 8.48 KB	File Size 20.9 KB
172 Triangles	428 Triangles
(*a*)	(*b*)

Figure 2.11   (*a*) Chord Height 1.0 and (*b*) chord height 0.1.

powdered metals. Most machines are fairly autonomous needing little human intervention. Build times vary depending on size and number of parts required.

2.2.5   Postprocessing

The final step in rapid prototyping is postprocessing. It essentially consists of part removal and cleaning and of postcuring and finishing. This step generally involves manual operations where an operator does the postprocessing with extreme care. Otherwise, the part may be damaged and may need to be prototyped again [7].

The tasks for postprocessing are different for different prototyping systems. Table 2.2 shows the necessary postprocessing tasks for some selected prototyping systems.

Part removal and cleaning refers to taking the prototype out of the prototyping machine and removing excess materials, including support materials, which may have remained on the part. As shown in Table 2.2, there is a need for cleaning with both the FDM machine and SLA machine. With SLA parts, excess resins residing in entrapped areas need to be removed. Likewise, excess powder for SLS parts and excess woodlike blocks of paper for LOM parts need to be removed.

Postcuring is a task that is usually only needed for SLA and SLS parts. In the SLA process, the laser scans each layer along the boundary and hatching lines only, resulting in side portions of the layers not being completely solidified. Postcuring is needed to complete the solidification process and to improve the mechanical properties of the prototype.

Postcuring is carried out in a specially designed apparatus using ultraviolet (UV) radiation. Optimizing the output wavelength of the postcuring apparatus

TABLE 2.2 Postprocessing Tasks for Various RP Systems

Postprocessing Tasks	Fused Deposition Modeling (FDM)	Solidscape	Stereolithography Apparatus (SLA)	Laminated Object Manufacturing (LOM)	Selective Laser Sintering (SLS)
1. Cleaning	✓	✓	✓	✓	✓
2. Postcuring	✓	✓	✓		✓
3. Finishing		✓	✓	✓	✓

determines the uniformity of postcuring with minimal temperature rise and maximum part accuracy. How long it takes to postcure a part depends on the part size. In general, time for postcure is less than that of part building. In any case, both postcuring and part building are unattended processes.

Part finishing is done once the part is postcured. For simple part application, part finishing involves basic cleaning such as sanding or machining to remove additional materials. Care should be taken when handling parts made of wax materials because the materials are very brittle.

2.3 PROBLEMS WITH STL FILE FORMAT

While the STL file format is meeting the needs of the industries that are using RP and while it is the de facto standard in RP industry, it has some inadequacies [2, 7]. Some of these inadequacies are due to the very nature of the STL file format as it does not contain topological data. Also, many CAD vendors use tessellation algorithms that are not robust. Consequently, they tend to create polygonal approximation models, which exhibit the following types of problems:

1. *Gaps (Cracks, Holes, Punctures) Indicating Missing Faces* When a solid model is converted into an STL file format, the solid model forms are replaced with a simplified mathematical form (triangles). However, if the simplified operation is not done properly, it introduces undesirable geometric anomalies, such as holes or gaps in the boundary surface. This problem is more prone to surfaces with large curvature. Such gaps are shown in Figure 2.12.

2. *Inconsistent Normals* In general, surface normals should be pointed outward. However, the normals of some surfaces could be flipped over, as shown in Figure 2.13, thus, becoming inconsistent with the outward orientation of the original surface.

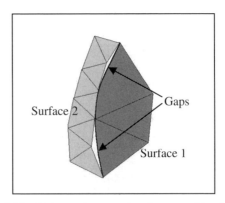

Figure 2.12 Polygonal approximations resulting in gaps.

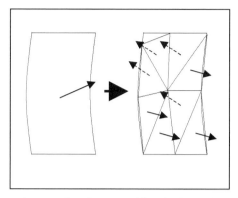

Figure 2.13 Polygonal approximations resulting in inconsistent and incorrect normals.

3. *Incorrect Normals* Sometimes, surface normals stored in the STL file are not the same as those computed from the vertices of the corresponding surfaces.

4. *Incorrect Intersections* Facets may sometimes intersect at locations other than their edges resulting in overlapping facets. (Fig. 2.14)

5. *Internal Wall Structure* Geometric algorithms are used for closing gaps in STL files. However, faulty geometric algorithms could generate internal walls and structures that can cause discontinuities in the solidification of the material (Fig. 2.15)

6. *Facet Degeneracy* Degeneration of facets occurs when they may not represent a finite area and consequently have no normals. Generally, there are two kinds of facet degeneracies: geometric degeneracy and topological degeneracy. A geometric degeneracy takes place when all the vertices of the facet are distinct and all the edges of the facet are collinear. A topological degeneracy takes place when two or more vertices of a facet coincide. Since it does not affect the geometry or the

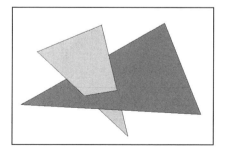

Figure 2.14 Polygonal approximations resulting in incorrect intersections.

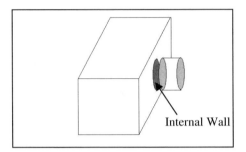

Figure 2.15   Polygonal approximations resulting in internal wall structures.

connectivity of the remaining facets, the faults can be discarded (Fig. 2.16)

7. *Inconsistencies*   Sometimes two STL files are combined to create a prototype. If these STL files were created using different tolerance values, it will lead to inconsistencies such as gaps.

2.4   OTHER TRANSLATORS

The problems and inefficiencies of the STL file format have prompted the search for alternate translators. Examples of some of these translators are IGES, HPGL, and computed tomography (CT) data.

1. *IGES File*   Initial graphics exchange specification is a common format to exchange graphics information between various CAD systems. It was initially developed and promoted by the then American National Standards In-

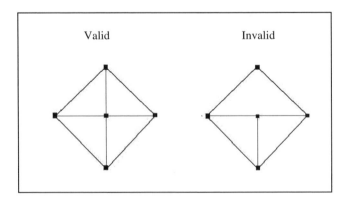

Figure 2.16   Polygonal approximations resulting in geometric degeneracy.

stitute (ANSI) in 1981. The IGES file can precisely represent both geometry and topological information for a CAD model. An IGES file contains information about surface modeling, constructive solid geometry (CSG) and boundary representation (B-rep). The Boolean operations for solid modeling such as *union, intersection,* and *difference* are also defined in the IGES file. It can precisely represent a CAD model by providing entities of points, lines, arcs, splines, NURBS surface, and solid elements. The primary advantage of IGES format is its widespread adoption and comprehensive coverage.

However, there are some disadvantages of the IGES format as it relates to its use as an RP format. These are:

1. The inclusion of redundant information for RP systems.
2. The algorithms for slicing IGES file are more complex than those for slicing STL file.
3. The support structures that are needed for some RP systems cannot be created using IGES format.

IGES is a very good interface standard for exchanging information between various CAD systems. It does, however, fall short of meeting the standards for RP system.

2. Hewlett–Packard Graphics Language (*HPGL*) *File*   HPGL has been the standard data format for graphic plotters for many years [1, 2]. Generally, data types are two-dimensional representing lines, circles, splines, text, and so forth. Many commercial CAD systems have the interface to output HPGL format, which is a two-dimensional geometry data format and does not require slicing. However, there are two major problems with the HPGL format. First, since HPGL is a 2D data format, the files are not appended, leaving hundreds of small files needing logical names and transformation. Second, all the required support structures must be generated in the CAD system and sliced in the same way.

3. CT Data   CT scan data is a new format used for medical imaging. The format has not been standardized yet. Formats are proprietary and vary from machine to machine. The CT scan generates data as a grid of three-dimensional points, where each point has a varying shade of gray indicating the density of body tissue present at that point. Data from CT scan are being regularly used to prototype skull, femur, knee, and other biomedical components on FDM, SLA, and other RP systems. The CT data essentially consist of raster images of the physical objects being imaged. It is used to produce models of human temporal bones.

Models using CT scan images can be made using CAD systems, STL interfacing, and direct interfacing. CT data is used to make human parts such as leg prostheses, which are used by doctors for implants. However, the main problem with CT image data is the complexity of the data and the need for a special interpreter to process this data.

2.5 CASE STUDY: DESIGNING AND PROTOTYPING A SPUR GEAR

This section describes a complete prototyping process using the FDM system. It also describes step-by-step procedures for making the prototype, as well as the hardware and software used for making the prototype.

2.5.1 Introduction

The spur gear has teeth parallel to the axis of rotation and is used to translate motion from one shaft to another. This type of gear-and-chain assembly is the same as the one used on motorcycles. A three-dimensional CAD file of a spur gear was created using AutoCAD and then rapid prototyped using QuickSlice software and an FDM 1650 rapid prototyping machine. The machine uses FDM technology to create parts in a relatively short period of time. The material used is ABS plastic filament, which is heated above its melting point and then extruded through a nozzle onto a build platform. When one layer is finished, the platform is lowered and a new layer is formed on top of it. The fact that FDM is an additive process allows it to make more complex parts than conventional subtractive CNC methods.

The AutoCAD software allowed for the part to be dimensioned and created accurately. In Figure 2.17 a two-dimensional drawing of the gear is shown to indicate the size of the part. The advantage of creating this part in AutoCAD is that the dimensions on the gear can be changed to allow for different gear ratios easily and quickly. This software allows almost any geometry to be created and then saved as an STL file.

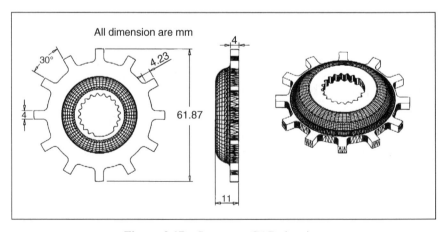

Figure 2.17 Spur gear CAD drawing.

2.5.2 Project Procedure

The fabrication process for this engineering prototype consisted of five steps: These include the CAD process, conversion to STL file, slicing the file, setup and build, and postprocessing. An image of the prototyping equipment used can be seen in Figure 2.18.

1. *CAD Process*
   a. A detailed CAD drawing of the part was created on AutoCAD (Version 14).
   b. The drawing created was a three-dimensional drawing and was made as a solid object.
   c. If there is more than one component to the model that is to be created, the "unity" function must be applied on AutoCAD. This is because the model should be one solid object in order to convert later to a stereolithography (STL) file.
   d. The CAD drawing was then converted into an STL file by choosing File and then Export commands.
   e. The file was now ready to be opened on the QuickSlice software.

2. *Part Preparation*
   a. The first step after opening the file was to properly orient the part to be created. Proper orientation can reduce support needed for the

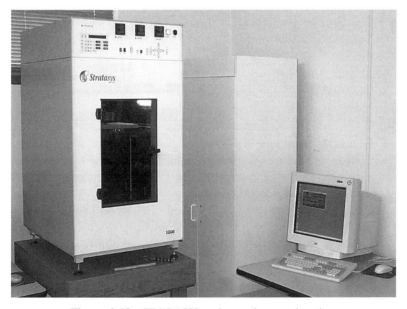

Figure 2.18 FDM 1650 and attendant workstation.

model and lessen valuable prototyping time. The orientation was set by choosing the tab at the left of the screen, which says Orient stl, and choosing the Rotate option. In this case, the gear was laid flat, which was the optimal position for this particular part.

b. Next, the part was sized in order to fill the build envelope. This can be done by selecting Orient .stl and choosing the Scale option.

c. Now that the part had been oriented and sized, it was ready to be sliced. The Slice tab was chosen and the Slice icon was selected. The software then did all the calculations necessary to slice the part. Once the part was sliced, only the bottom layer could be viewed, so the All icon was chosen. This allows for all the layers to be viewed at the same time, the model should be all red, if not there is a discontinuity in the structure of the model. Therefore, another .stl file must be created to fix the problem.

d. If the model is completely red, then support can now be added to the model. The Create Support icon was chosen and the software analyzed the part and added support where necessary. After the necessary support was added, the Create Base icon was selected and a base was added to the part. This gives the machine a flat surface to lay down the ABS plastic.

e. The next, and one of the most crucial steps, was creating the roads for the part. This is the path the extrusion head will follow to create the part. The Create Roads icon was selected. The software then created the appropriate paths.

f. The final step was to save the created file to .sml format. This file format allows the machines to read the information contained in the file.

3. *Set Up and Build*

a. The FDM machine was operating at an overall temperature of 68°C (envelope temperature). The model extrusion head was operated at 270°C and the support material head was operating at 265°C.

b. The FDM Send icon was selected and a file window popped up. The appropriate file was chosen from this window. The machine was now receiving the .sml file information. The Pause button on the machine was blinking at this time.

c. A 10 × 10-in. substrate was placed into the building chamber. The Pause button was pushed. The machine now reset itself to the home position. Now the part origin was set and the nozzle tip was placed slightly into the substrate. The Pause button was pressed again and the machine began making the part.

4. *Postprocessing*

a. Support removal was accomplished with the use of simple hand tools (e.g., spatula, knife, etc.).

   b. Sand paper was also used to completely remove the base material from the base.

5. *Conclusions*   This project allowed a part to be designed and then created using RP technology. An image of the final part produced can be seen in Figure 2.19. A project such as this can provide an invaluable experience to a group of young engineers in order to understand the process of making a design, and having a prototype to test before full production of the part begins. This is important when entering industry, where knowledge of cutting edge technology such as RP can give one an advantage over others.

    This experience can also help prove the efficiency and flexibility of a CAM process, which involves rapid prototyping. The machine can produce over eight different parts with completely different geometry in less than 3 weeks. This is actually remarkable for a machine that takes up no more room than a closet. There are size and material limitations, but the advantages definitely outweigh the disadvantages, which are limited part size and less than optimum surface finish.

2.6 SUMMARY

In this chapter the basic concepts of the rapid prototyping process were introduced and discussed. Rapid prototyping is essentially a part of *automated fabrication technology* that allows one to make three-dimensional parts from a digital design. Fabrication process is generally classified as subtractive, additive, and formative or even hybrid. Rapid prototyping is an additive process. All rapid prototyping processes basically consist of five steps: creation of solid model, conversion of solid model to stereolithography file format, slicing the file using a software package, actually making the prototype, and post-

Figure 2.19   Prototype of the spur gear.

processing of the prototype. The most common file format commonly used in industries and academia is the STL format, developed by the Albert Consulting Group for 3D Systems. Although the file format is not perfect, it is still being used as de facto standard for rapid prototyping industry.

PROBLEMS

2.1. What is a rapid prototyping system? Describe briefly a system showing all its components.

2.2. What are the purposes of data creation?

2.3. What is the most common file format used for RP?

2.4. What are the advantages of the STL file format?

2.5. What are the shortcomings (problems) of the STL file format?

2.6. Mention several translators that can be used in place of STL format. What are some drawbacks or limitations of these?

2.7. What is postprocessing? Do you need postprocessing for all RP systems?

2.8. Laboratory project: Design and prototype a product.
   a. Use a standard CAD to design the part shown in Figure P2.1.

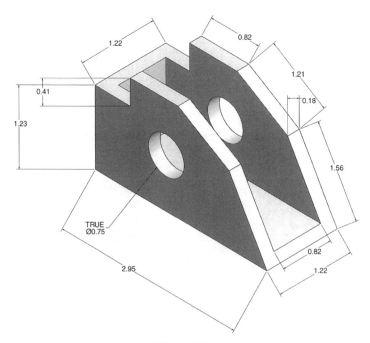

Figure P2.1

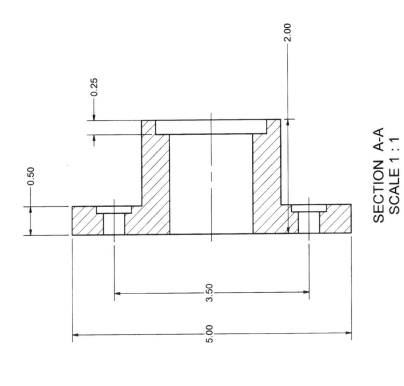

0.25

2.00

0.50

3.50

5.00

SECTION A-A
SCALE 1 : 1

Figure P2.2

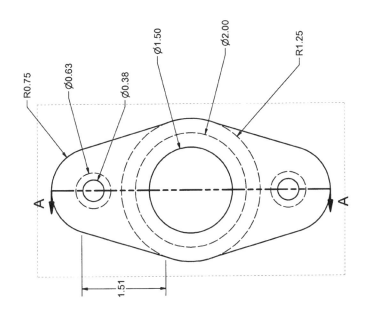

R0.75

Ø0.63

Ø0.38

Ø1.50

Ø2.00

R1.25

A

A

1.51

55

    b. Convert the drawing file into an STL file using conversion rules described in the text.

    c. Use an available RP system (FDM, SLA, LOM, SLC, or 3DP) to prototype the part.

    d. Use postprocessing techniques to complete the prototyping cycle.

2.9. For the object as shown in Figure P2.2:

    a. Draw the part using Pro/Engineer or a similar CAD/CAM system.

    b. Create STL binary file with chord height of 1.0 and 0.1.

    c. Create STL ASCII file with chord height of 1.0 and 0.1.

    d. Discuss the changes that take place when chord height is varied (file size, number of triangles, etc.).

REFERENCES

1. M. Burns, *Automated Fabrication: Improving Productivity in Manufacturing,* Prentice-Hall, Englewood Cliffs, NJ, 1993.

2. P. Jacobs, *Rapid Prototyping and Manufacturing & Fundamentals of Stereolithography,* SME Publications, Hong Kong, 1992.

3. S. Ashley, Rapid Prototyping Systems—Special Report, *Mechanical Engineering,* 34–43, 1991.

4. L. Wood, *Rapid Automated Prototyping: An Introduction,* Industrial Press, New York, 1993.

5. H. Bohn, File Format Requirements for the Rapid Prototyping Technologies of Tomorrow, International Conference on Manufacturing Automation Proceedings, Hong Kong, 1997.

6. J. Miller, Fundamentals of Rapid Prototyping and Applications in Manufacturing, Westec Conference, Los Angeles, CA, 1996.

7. C. Kai and K. Fai, *Rapid Prototyping: Principles and Applications in Manufacturing,* Wiley, New York, 1997.

3

LIQUID-BASED RP SYSTEMS

Rapid prototyping systems can be classified in a variety of ways depending on the physics of the process, the source of energy, type of material, size of prototypes, and the like. For the purposes of this book, a classification scheme based on the initial form of the material is presented. Accordingly, all RP systems can be classified as solid-based, liquid-based, or powder-based systems.

In this chapter, we shall study the liquid-based RP systems that include 3D Systems' stereolithography process, D-MEC's solid creation system (SCS), and CMET's solid object UV laser process. For each system, information on company profile, the principle, the process, the products, advantages and disadvantages, and application is provided.

3.1 CLASSIFICATION OF RP SYSTEMS

The most widely accepted means of categorizing RP systems in the industry is by determination of the initial form of the raw material. This text will follow the accepted norm. Accordingly, all RP systems can be categorized into solid-based, liquid-based, and powder-based systems [1].

3.1.1 Liquid-Based Systems

Liquid-based RP systems begin with the build material in the liquid state. The liquid is converted into a solid state through a curing process. The following RP systems fall into this category:

1. 3D Systems stereolithography spparatus (SLA)
2. D-MEC's solid creation system (SCS)
3. CMET solid object UV laser plotter
4. EOS's stereos system
5. Meiko's rapid prototyping system

3.1.2   Solid-Based Systems

Solid-based RP systems begin with the build material in the solid state. The solid form may include the material in the form of a wire, a roll, laminates, or pallets. The following RP systems fall into this category:

1. Stratasys' fused deposition modeling (FDM)
2. Helisys' laminated object modeling (LOM)
3. Solidscape, Inc.'s ModelMaker-6B
4. 3D System's multijet modeling (MJM) systems
5. Kira's selective adhesive and hot pass (SAHP) system
6. IBM's rapid prototype system (RPS)
7. Laser-engineered net shaping (LENS)
8. Solidica

3.1.3   Powder-Based Systems

While powder is considered solid, a special category of powder-based system has been intentionally created outside the solid-based systems to represent powder in granular form. The following RP systems fall into this category:

1. 3D Systems' selective laser sintering (SLS)
2. Soligen's direct shell production casting (DSPC)
3. Fraunhofer's multiphase jet solidification (MJS)
4. MIT's 3D printing
5. EOS's laser sintering

3.2   3D SYSTEMS' STEREOLITHOGRAPHY APPARATUS (SLA)

3.2.1   Company Profile

3D Systems, Inc. develops, manufactures, and markets stereolithography equipment designed to produce solid, three-dimensional plastic parts directly from CAD/CAM data. The company has quickly grown from pioneer to proven leader in rapid prototyping with hundreds of installations worldwide.

Based in California,* 3D Systems, founded by Charles Hull, the inventor of stereolithography, serves customers virtually worldwide, with offices in the United States, United Kingdom, France, Italy, Hong Kong, and Japan and employs more than 300 people worldwide.

A tightly integrated combination of hardware, software, materials, and process gives 3D Systems one of the broadest ranges of solid imaging solutions in the world. The comprehensive range of products consists of the SLA (stereolithography) product line, the SLS (selective laser sintering) product line, the MJM (multijet modeling) product line, and a broad range of materials for all of 3D Systems Solid Imaging Products.

3.2.2 Details of Stereolithography ("SL") Process

The SL process is fundamentally based on parts that are built from a photocurable liquid resin that solidifies when sufficiently exposed to a laser beam (basically undergoing a photopolymerization process) that scans across the surface of the resin. The building is done layer by layer, each layer being scanned by the optical scanning system and controlled by an elevation mechanism that lowers at the completion of each layer. The details that stereolithography incorporates, such as the stereolithography process, the STL file format, C-Slice processing to STL files, thermoplastic resin, the SL manufacturing process, and quantitative analysis of laser-based manufacturing, are described in this section [2].

Stereolithography Process   The stereolithographic process generates a part in the following manner:

1. A photosensitive polymer that solidifies when exposed to a UV light source is maintained in a liquid state.
2. A platform that can be elevated is located just one layer of thickness below the surface of the liquid polymer.
3. The UV laser scans the polymer layer above the platform to solidify the polymer and gives it the shape of the corresponding cross section. This step starts with the bottom cross section of the part.
4. The platform is lowered into the polymer bath to one layer thickness to allow the liquid polymer to be swept over the part to begin the next layer.
5. The process is repeated until the top layer of the part is generated.

---

*3D Systems, Inc. corporate headquarters: 26081 Avenue Hall, Valencia, CA 91355. Toll-free number: (888) 337-9786; corporate phone number: (661) 295-5600; fax: (661) 294-8406; web: www.3Dsystems.com.

6. Postcuring is performed to solidify the part completely. This is required because some liquid regions can remain in each layer. Because the laser beam has finite size, the scanning on each layer is analogous to filling a shape with a fine color pen.

The stereolithography process of generating a part described in the steps above is illustrated in Figure 3.1. A picture of a stereolithography machine is shown in Figure 3.2.

STL File Format The STL file format was introduced by 3D Systems in 1987, and, since then, it has become the de facto standard of the industry, even though other slice file formats are available. As discussed in Chapter 2, the STL format translates the CAD model into triangles just like the hexagons and pentagons on the surface of a soccer ball.

The STL file format consists of (1) a header, (2) the number of triangles, and (3) a list of triangle description by vertices and the normal vector to the triangles [3]. The structure of a standard STL file format is shown in Table 3.1.

Each triangle of the STL file contains 50 bytes; 12 floating-point numbers plus an unsigned integer or $(12 \times 4) + 2 = 50$. The size of the STL file is 50 times the number of triangles, plus 84 bytes for the header and the triangle count [1]. Therefore, a 10,000-triangle object requires 500,084 bytes in size.

There are two very important rules that govern the proper processing of the STL file:

Rule 1   Using triangle vertex ordering, the interior or exterior surfaces are identified. For example, the right-hand counterclockwise rule, or "ccw rule," is a corkscrew acting outward on a soccer ball, ordering the vertices and the normal vector (Fig. 3.3).

Rule 2   The vertex-to-vertex rule requires that the vertices of adjacent triangles link to the neighboring triangle and that no vertices meet a neighboring edge (Fig. 3.4).

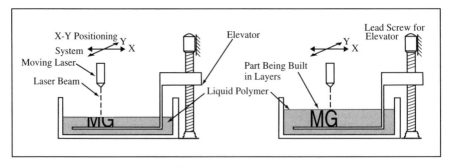

Figure 3.1   Stereolithographic process generating a part.

Figure 3.2 3D Systems SLA 3500 Systems. (Courtesy of 3D Systems, Inc.)

TABLE 3.1 STL File Format

Entity	Described By
Header	80 bytes
Number of triangles	Unsigned long integer (4 bytes)
For each tessellation triangle (50 bytes of information)	See below
Normal vector I	Floating-point integer (4 bytes)
Normal vector J	Floating-point integer (4 bytes)
Normal vector K	Floating-point integer (4 bytes)
First vertex X	Floating-point integer (4 bytes)
First vertex Y	Floating-point integer (4 bytes)
First vertex Z	Floating-point integer (4 bytes)
Second vertex X	Floating-point integer (4 bytes)
Second vertex Y	Floating-point integer (4 bytes)
Second vertex Z	Floating-point integer (4 bytes)
Third vertex X	Floating-point integer (4 bytes)
Third vertex Y	Floating-point integer (4 bytes)
Third vertex Z	Floating-point integer (4 bytes)
Attribute	Unsigned integer (2 bytes)

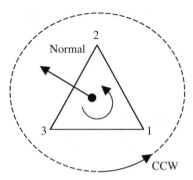

Figure 3.3   Rules for tessellation.

C-Slice Processing to STL Files   Slicing of an STL file is necessary to produce the vector data necessary for scanning. The slicing process consists of sorting the STL triangles into "z values," finding the boundary segments, creating boundary polylines, applying edge compensations, and the like. The flowchart for the entire slicing process [1] is shown in Figure 3.5.

Thermoplastic Resin   Liquid photopolymers can be solidified by electromagnetic radiation whose wavelengths may include γ-rays, X-rays, UV, and visible range or electron beam (EB) [1]. Most RP systems including SLA systems are curable in the UV range. The photocurable liquid was initially developed for printing and for using as furniture lacquer/sealant. The UV curing process was developed to avoid liquid that had carcinogenic solvents. Lasers provide more direct energy, which allowed 3D Systems to invent the stereolithography process. The availability of powerful computers that could create tessellated solid models also helped develop the SL process.

Photopolymerization is defined as the linking of small molecules (monomers) into larger molecules (polymers). Typical polymers consist of hundreds of thousands of monomers that can have molecular weights up to thousands of grams per mole [2]. Vinyl monomers have a carbon–carbon (C=C) double

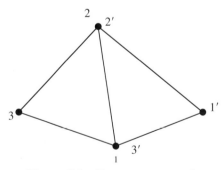

Figure 3.4   Vertex-to-vertex rule.

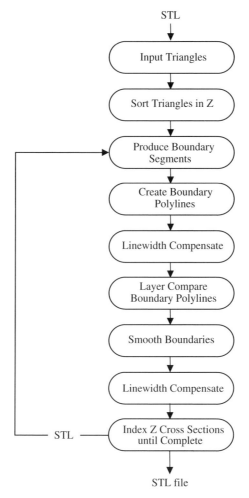

Figure 3.5   STL data process flow.

bond attached to complex groups donated by the free radical R. In the original resin, the monomer groups are only weakly connected to their neighbors by van der Waals bonds. As these bonds are acted on by lasers, the C=C bonds break. The broken monomer groups then connect to each other, forming long chains as shown in Table 3.2.

The bonding between such chains creates the following three key effects:

1. The liquid bonds into a solid.
2. The density increases.
3. The shear strength increases.

TABLE 3.2   Polymerization Process

Weak Van der Waals Bonds Between Adjacent Chains	Strong Covalent Bonds Along Chains
$\begin{array}{c} R \\ | \\ H_2C = CH \\ H_2C = CH \\ | \\ R \end{array}$	$\begin{array}{c} -CH_2-CH-CH_2-CH- \\ \quad\quad | \quad\quad\quad | \\ \quad\quad R \quad\quad\quad R \end{array}$

Although the original vinyl monomers are already cross-linked, they get much more strength from the formation of covalent bonds in the long chains.

SL Manufacturing Process   The process known as stereolithography is a form of "layered manufacturing" and has become an increasingly popular process for rapid prototyping. The components of this manufacturing process consists of a vat of liquid photocurable monomers, a computer-controlled platform that only moves vertically in the vat, and a laser beam above that shines down on the surface of the vat and that only moves horizontally [3].

Each individual layer initially requires boundaries to be traced out by the laser, after which hatching or weaving patterns are projected onto the entire area. Hatched areas are then filled in, promoting the gelling and solidification stage. Figure 3.6 shows the bordering, hatching, and filling process. Subsequent layers follow this process until the part is complete [3]. *Note:* The steps described below are for an SL 250 system at the time of this writing.

Step 1. Script Preparation   The part being produced requires instructions on desired accuracy, Zephyr blade sweeping times, and "z-wait" times. The layer thickness typically averages 100 μm (0.004 in.) for the build layer. However, the complete range capability is from 50 to 200 μm (0.002 to 0.008 in.). This information, input by the operator, is called the script.

Step 2. Leveling and Laser Calibration   The SL resins are continuously shrinking. Therefore, in order to achieve the proper level, a sensor must

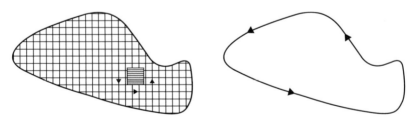

Figure 3.6   Bordering, hatching, and filling for one square.

be installed to sense the liquid level in the vat. If the liquid level is not at the desired position for the beginning of the run, a plunge mechanism adjusts the fluid level by fluid displacement. Laser calibration incorporates reflective "eyes" in order to check the position before each layer is done.

Step 3. Initial Supports   The supports for a part can be viewed as small feet, similar to those on a sofa or table. The process of lifting the part off the floor of the platform requires these supports so that the Zephyr recoater blade does not hit the platform. The supports are also needed so that it is easier to remove the finished part and in order to compensate for any platform distortion. Lastly, internal supports are required for any over-hanging structures.

For the first layer, the staging platform is moved automatically to about 5 mm (0.2 in.) for the SL 250 system. This dip allows the viscous fluid to flow over the surface of the first layer of the supports. The elevator positions itself 100 μm (0.004 in.) below the surface. Before laser curing, it is normal to wait 5 sec. This creates the second layer of the support and so on until the supporting stubs have achieved the desired height. The standard height of the supports is 7.6 mm (0.300 in.).

Step 4. Part Creation   For the actual part the procedure is somewhat differ-ent. Once the supports have been created, the first bottom surface of the part is generated by the "bordering, hatching, and filling" described earlier. The elevator proceeds with a descent of 100 μm (0.004 in.) and waits for about 0 to 10 sec. The operator has chosen this time during the script creation. The laser has begun the polymerization process and takes about 10 to 20 sec for the polymerization to occur and harden in order for the next layer to be built on top. After the first layer is hardened enough, the Zephyr blade sweeps over the surface and deposits the 100-μm (0.004-in.) layer of liquid for the second polymerization.

Step 5. Zephyr Blade Sweeping   The squeegeelike Zephyr blade is long, has a hollow cavity between two adjacent blades, and is under the influence of a slight vacuum. The vacuum draws liquid into the bottom blade and ensures uniformly deposited liquid layers. As the Zephyr blade sweeps across the vat, it removes as well as fills in areas in order to achieve a smooth surface. The sweep takes about 5 sec unless a hollowlike part is being made where the viscous fluid inside the hollow takes longer to follow the blade. The sweep gives a uniform thin layer, although there is a ten-dency for resin to adhere to the blade.

Step 6. Relaxing the Fluid   After all the adjustments and sweeping, a crease exists around the edge of the part at the solid–liquid interface. A Z-wait time of about 0 to 10 sec is required to allow the fluid to flatten and the resin surface to smooth.

Step 7. Filling   Toward the end of the process, a more intense hatching may be desirable on the top surface of the part. Closely spaced vectors cause more intense solidification structures on up-facing surfaces.

Step 8. Finalizing Final steps include:
   Draining excess resin from cavities
   Cleaning and rinsing with solvents
   Removing supports
   Hand sanding and polishing
   Postcuring in broad-spectrum UV light

Quantitative Analysis of Laser-Based Manufacturing Stereolithography, selective laser sintering, and a few other RP processes use the power of laser to control the solidification and accuracy of prototypes made from resin or powder. RP is a layer-by-layer process in which each layer adheres to the next layer to form a solid. The bottom layer of the prototype is made first, followed by the successive upper layers, all adhered to each other. The energy of the laser at a depth z is of much interest to the researchers. Lasers are capable of causing *polymerization by irradiance,* compared to regular arc lamps. However, the energy of a laser is not constant as it travels through the thermoplastic resin or powder, rather it decays exponentially following the Beer–Lambert exponential law of absorption as shown below in the equation [3]

$$H_{(x,y,z)} = H_{(x,y,0)} \exp\left(-\frac{z}{D_p}\right) \tag{3.1}$$

where $H_{(x,y,z)}$ = exposure at position x, y, z
   $H_{(x,y,0)}$ = exposure at bottom x, y, 0
   z = any height
   D_p = resin constant, defined by depth of a particular resin that results in a reduction of irradiance level to $1/e(=1/2.718)$ of
   the H_0 level on the surface (Fig. 3.7).

It means that at a depth of $z = D_p$, the irradiance is $1/2.718 = 37\%$ of H_0. Typical values of SLA 250 system are as follows [2]:

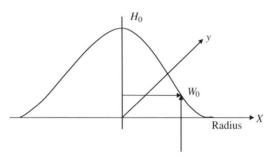

Figure 3.7 Gaussian decay of laser energy across the surface.

$$\text{Nominal laser power} = (P_L) = 15 \text{ mW}$$

$$\text{Central spot size} = (2W_0) = 0.25 \text{ mm}$$

For the entire spectrum, the Gaussian irradiance curve controls the physics, and like any other point source of light it decays from the center. The decay of the laser energy across the surface, as opposed to down through the surface, is given by the equation

$$H_{(x,y,0)} = H_{(r,0)} = H_0 \exp\left(-\frac{(2)(r)^2}{W_0^2} \right) \tag{3.2}$$

where $W_0 = 1/e^2$, Gaussian half-width (Fig. 3.8). Thus, at $r = W_0$,

$$H = H_0 e^{-2} = 0.135 H_0$$

It also follows that

$$H_{\text{average}} = \frac{P_L}{\pi W_0^2}$$

$$= \frac{15}{\pi(0.25/2)^2} \text{ mW} \tag{3.3}$$

and thus

$$H_{\text{average}} = 30.56 \text{ W/cm}^2 \tag{3.4}$$

Figure 3.8   Viper SLA System. (Courtesy of 3D Systems, Inc.)

If the scanning speed of the laser is 200 mm/sec, the laser's exposure time (t_e) on a given area is given by

$$t_e = \frac{2W_0}{V} = \frac{2 \times 0.125}{200} = 1.25 \text{ msec} \tag{3.5}$$

The average energy density of the laser exposure is given by

$$E_{(average)} = H_{average} \times t_e$$

$$= 30.50 \frac{\text{W}}{\text{cm}^2} \times 1.25 \text{ msec}$$

$$= 38.2 \frac{\text{mJ}}{\text{cm}^2} \tag{3.6}$$

We shall now proceed to calculate the polymerization ability from the laser's photon flux. Using Planck's equation, the photon energy is found to be

$$E_{(photon)} = \frac{hc}{\lambda} \tag{3.7}$$

where E = photon energy
  h = Planck's constant
  c = velocity of light
  λ = laser wavelength

So, $$E_{photon} = \frac{(6.62 \times 10^{-34} \text{ J} \cdot \text{s}) \times (3 \times 10^{10} \text{ cm/s})}{3.25 \times 10^{-5} \text{ cm}}$$

$$= 6.1 \times 10^{-19} \text{ J/photon} \tag{3.8}$$

Assuming that N_{ph} is the number of photons per square centimeter hitting the resin surface,

$$N_{ph} = \frac{E_{average}}{E_{photons}} = \left[\frac{38.2 \text{ mJ/cm}^2}{6.1 \times 10^{-19} \text{ J/photon}} \right] 6.3 \times 10^{16} \frac{\text{photons}}{\text{cm}^2} \tag{3.9}$$

This flux of photons penetrates into the resin and acts on the polymer chains to cause polymerization. Even with photochemical efficiency of 50%, the [C=C] double bonds will polymerize to [—C—C—] single bonds.

3.2.3 SLA Systems

3D Systems introduced the original SLA 1, SLA 190, SLA 250 series, then SLA 500, SLA 3500, SLA 5000, SLA 7000 machines, and the most recent

Viper SLA system. More recent SLA systems greatly enhance the application capabilities compared to earlier SLA systems technology. A large variety of photocurable materials are also available for use with SLA systems. We shall briefly discuss some of the newest systems in this section.

Viper SLA System   3D Systems' new Viper SLA system is the first solid imaging system to combine standard and high-resolution part building in the same system. One of the two modes can be chosen depending on the needs:

1. *Standard Resolution Mode*   For the best balance of build speed and part resolution
2. *High-Resolution* (*HR*) *Mode*   For ultra-detailed small parts and features

All features form a carefully integrated digital signal processor (DSP) controlled by a high-speed scanning system with a single, solid-state laser that delivers 100 mW of available power.

The applications of this system are:

Small to medium-sized concept and communications models

Small to medium-sized prototypes

Patterns for injection molding dies and patterns for investment casting

Precisions builds of extremely detailed parts

Parts with extremely fine details

Advantages of the system:

High levels of detail and accuracy

High-resolution capability

Cost saving compared to other RP techniques

High productivity

SLA 7000 System   The SLA 7000 system is three times faster, on average, than the next fastest solid imaging system from 3D Systems. But raw speed is just the beginning. The SLA 7000 system's 0.0254 mm layer thickness yields a smooth finish that results in far less postprocessing time. Highly reliable, the fifth-generation design includes fewer parts, a low-vibration optical system, and revolutionary dual-spot laser technology—reducing downtimes. Applications for this system are:

Limited production runs

Rapid tooling prototyping

Master patterns for investment casting

Form, fit, and function testing

Concept communication modeling

SLA 5000 System   The SLA 5000 system incorporates patented Smart-Sweep™ technology eliminating unnecessary sweeper motions. A 0.05-mm build layer creates a smooth finish that requires far less postprocessing time. At 508 × 508 × 584 mm the build envelope invites larger single parts or multiple smaller parts at once. Solid-state laser technology delivers up to 35% more power over the SLA 3500 system. Applications for this system are:

   Prototypes for design verification and testing
   Patterns for casting and molding
   Tools for preproduction tooling
   Parts for manufacturing aids, quote enhancement, and limited production
      runs

SLA 3500 System   The SLA 3500 system runs patented SmartSweep technology on a 0.05-mm build layer creating accurate parts. The SLA 3500 is up to 2.5 times faster than the SLA 250 system. The SLA 3500 system is compatible with a range of 3D Systems' stereolithography materials. Applications for this system are:

   Patterns for casting and molding
   Tools for preproduction tooling
   Parts for manufacturing aids, quote enhancement, and limited production
      runs
   Testing under heat
   Complex fit tests
   Flow visualization
   Wind tunnel testing
   Optical stress analysis

SLA 500 System   The SLA 500 system part accuracy and part building reaches speeds of up to 34% faster than the SLA 350 series and up to 73% over the SLA 250 series. A 500 mm × 500 mm × 584 mm (20 in. × 20 in. × 23 in.) build envelope invites larger single parts or multiple smaller parts at once. The SLA 500 systems incorporates Zephyr recoating technology and higher powered laser. Applications for this system are:

   Prototypes for casting and molding
   Tools for preproduction tooling
   Parts for manufacturing aids, quote enhancement, and limited production
      runs

Table 3.3 shows the product specifications for various SLA models.

TABLE 3.3 SLA Machine Specifications by Model

Parameter	SLA-250	SLA-500	SLA-3500	SLA-5000	SLA-7000	Viper si2
Laser type	He–Cd	Argon ion	YVO_4	YVO_4	YVO_4	YVO_4
Laser power (mW)	7.5	132–264	160	216	800	100
Spot size (mm)	0.2–0.29	0.2–0.3	0.2–0.3	0.2–0.3	0.02–0.03	0.02–0.03
Elevator vertical resolution (mm)	0.0025	0.00177	0.00177	0.00177	0.00100	0.00250
Vat capacity (liter)	29.5	253.6	99.30	253.6	253.6	32.21
Work volume, XYZ (mm × mm × mm)	250 × 250 × 250	500 × 500 × 584	350 × 350 × 400	508 × 508 × 584	508 × 508 × 600	250 × 250 × 250
Max part weight (kg)	9.1	68.04	56.80	68.04	68.04	9.100

3.2.4   Advantages and Disadvantages

Advantages

- Market share and industry presence quickly achievable
- Capable of high detail and thin walls
- Good surface finish
- Continuous process improvement by 3D Systems

Disadvantages

- Postcuring required
- Some warpage, shrinkage, and curl due to phase change
- Limited material (photopolymer)
- Support structures needed
- Standard lab safety requirement for handling of non-post-cured materials

3.2.5   Applications and Uses

Stereolithography apparatus systems provide manufacturers cost-justifiable methods for reducing time to market, lowering production costs, gaining greater control of their design process, and improving product design. The range of applications include:

1. Models for conceptualization, packaging, and presentation
2. Prototypes for design, analysis, verification, and functional testing
3. Masters for prototype tooling and low-volume production tooling
4. Patterns for investment casting, sand casting, and modeling
5. Tools for fixture and tooling design and production

3.3   D-MEC'S SOLID CREATION SYSTEM

3.3.1   Company Profile

D-MEC is a Japanese company that is involved in many business areas related to the molding system that has established its position as an RP maker, known as solid creation system (SCS). This system is based on stereolithography systems in which 3D CAD data, laser lights, and UV-curable resins are combined to create a solid. The system was jointly developed by Sony, JSR, and D-MEC. D-MEC is involved with D-MEC modeling, sales, and services of SCS, design of direct mold, mold modeling, and commissioned manufacturing of injection-molded models, and sales of UV-curable resin (DESOLITE). The products of D-MEC are used in automobiles, office automation devices, domestic appliances, precision machines, medical equipment, and so forth [4].

3.3.2 Solid Creation System

The SCS is a leading-edge molding/RP system using light in which UV-curable resin (DESOLITE) is laminate-formed using a laser beam according to 3D data. This technique creates real models with the exact same dimensions as those of a 3D CAD model in a shorter period of time. The SCS has been a jointly developed project. Sony has been responsible for the hardware and software, while JSR (Japan Synthetic Rubber) has been responsible for the developed UV-curable resin, and D-MEC for the forming and applied technology.

The system uses an ultraviolet laser to draw cross-sectional patterns on the resin solution surface. The ultraviolet beam chemically reacts with the resin solution in order to form a thin solid layer. The beam is swept across the fluid in order to build up continual thin layers that form a 3D model. See Figure 3.9.

3.3.3 Products

The SCS 2000 is suitable for cases where less sophistication is required. Table 3.4 gives some of the specifications of the SCS 2000.

The SCS 2000 is shown in Figure 3.10.

The SCS machines range from low speed and small sized to high speed and large sized. Some of the general specifications for each of the machines is shown in Table 3.5.

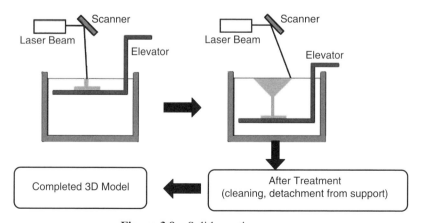

Figure 3.9 Solid creation process.

TABLE 3.4 Specifications of the SCS 2000

Optical Unit	
Laser	Semiconductor excitation solid-state laser
Modulator	Superspeed modulator
Deflector	Galvanometer mirror type (with defocus via sweep and straightening function)
Modeling range	600 × 500 × 500 mm
Spot size	0.15–0.8 (auto variation)
Speed	Max. 10 m/s
Lamination thickness	0.05–0.4 mm
Power Supply and Machinery	
Tank capacity	2.65 liter
Temperature control	50°C or less
Machine size	1650 × 1400 × 1820 mm
Machine weight (including resin)	1290 kg
Power supply	AC100V 30A
Cooling water	Not used
Control Unit	
Controller	Personal computer OS: Windows NT

Figure 3.10 SCS-2000.

TABLE 3.5 SCS Machine Specifications

Products	Description	Molding Range (mm)	Laser	Price (US $)
SCS-100HD	High-precision and high-definition machine	300 × 300 × 270	He–Cd laser	425,000
SCS-300P	Small-sized machine for general purpose	300 × 300 × 300	0.2W, semiconductor excitation solid-state	208,334
SCS-2000	Medium-sized machine for general purpose	600 × 500 × 500	0.2W, semiconductor excitation solid-state	416,667
SCS-8000	High-power and high-speed machine	600 × 500 × 500	1.8W, semiconductor excitation solid-state	541,667
SCS-8000 Duet Scan	Super-high-power high-speed machine	600 × 500 × 500	5W, semiconductor excitation solid-state	1,000,000
SCS-3000	Large-sized machine	1000 × 800 × 500	1.8W, semiconductor excitation solid-state	675,000

3.3.4   D-MEC Modeling

D-MEC utilizes Materialise's Magics (Fig. 3.11*a*) data processing software. The software package incorporates STL measurement/correction, support creation, and D-MEC's slice data format (SCDB) translation. For purposes of browsing and editing slice data, Solid Ware (Fig. 3.11*b*) software has been added, which includes D-MEC's extensive experience in the slice data-editing field.

Advantages and Disadvantages

Advantages

- No poisonous fumes or offensive odor during the process
- High-speed scanning for quick production of large models
- Large tank size enables manufacturing of full-scale prototypes
- High accuracy (0.04 mm repeatability) model may be produced

Disadvantages

- Costly maintenance of the laser
- Lack of variety in materials used

Applications and Uses   Sony's SCS machines provide manufacturers with the ability of optimum design in its products and to perform important functionality tests on the prototype. The range of applications include:

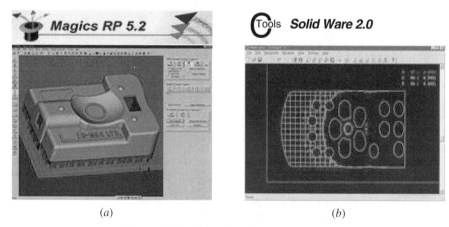

(*a*)   (*b*)

Figure 3.11   D-MEC software package.

1. Product design conceptualization
2. Design study and test samples of new product
3. Parts created without modification needs in small lot production
4. Mold tool and master investment casting and similar processes

3.4   CMET SOLID OBJECT UV LASER PLOTTER

3.4.1   Company Profile

In April of 2001, Teijin Seiki acquired and merged with NTT Data CMET to form a new company called CMET Inc. Teijin Seki was already a diversified industrial technology company in Japan that was promoting and selling Soliform rapid prototyping machines. Its Soliform system is based on the SOMOS (solid modeling system) laser-curing process developed by DuPont Imaging Systems. The merger of NTT Data CMET and Teijin Seiki has reduced the number of RP companies promoting stereolithography processes in Japan. However, other small to medium-sized companies such as Autostrade, Denken, and Meiko have joined the RP industry [5].

3.4.2   SOUP Process

The solid object UV laser plotter (SOUP) process applies the principles of stereolithography to prototype parts. Most machines use He–Cd lasers and photocurable epoxy resins. The SOUP process contains three main steps that are shown below [1]:

1. *Obtaining 3D Model from CAD System*   A 3D model, usually a solid model of the part is created with a commercial CAD system. Afterwards, the 3D data of the part is generated.
2. *Processing Data with SOUPware*   SOUP's software, SOUPware, is used to edit CAD data, fix defects such as gaps and overlaps, and slice the model into cross sections. Then SOUPware generates the necessary corresponding SOUP machine data.
3. *Making Models with SOUP Units*   The resin is solidified using the laser, into the cross-sectional data provided from SOUPware. Next the elevator is lowered and the liquid covers the top layer of the part, which is then recoated and prepared for the next layer. This step is repeated until the whole part is created.

3.4.3   SOUP Machines

Aside from the SOUP machine models shown in Table 3.6, the SOUP process incorporates the usage of hardware such as communications, laser, shutter/filter, scanner and elevator controller [5]. SOUPware is a multiuser and multi-

TABLE 3.6 SOUP's Models and Specifications by Model

Parameter	SOUP600GA		SOUP1000GA
Laser type	He–Cd	AR	Ar
Laser power (mW)	35–45	400	400
Spot size (mm)	0.2		0.2
XY sweep speed (m/s)	6		6
Scan resolution (mm)	±0.05		±0.05
Work volume, XYZ (mm × mm × mm)	600 × 600 × 500		1000 × 800 × 500
Minimum layer thickness (mm)	0.05–0.3		0.05–0.3
Size of unit, XYZ (m × m × m)	1.7 × 1.09 × 1.85		2.20 × 1.50 × 2.50
Data control unit	SUN WS		SUN WS
Power supply	400 V_{AC}, 3 phase, 25–40 A		
	(additonal 60 A for Ar laser)		
Price (US $)	666,400		836,000

machine control software. SOUPware has the ability to run simulations, perform editing for the repair of defects, 3D offset, loop scan for the filling of the area between the outlines, detachable structure, and automatic support generation. Because the SOUP system is based on the laser lithography technology, which emphasizes faster scanning, the laser beam diameter is very small (0.2 mm max).

3.4.4 Advantages and Disadvantages

Advantages

- Can create parts with special and complex structures
- Recoating system with more accurate Z layers
- Shorter build time
- Machines have a replaceable resin vat
- Software system allow for real-time processing

Disadvantage

- Maintenance of expensive lasers

3.5 SUMMARY

In this chapter, the principles and operation of liquid-based rapid prototyping systems are introduced and discussed. Although there are a few other liquid-based systems in the world, we have described the most commonly used systems, such as the 3D Systems stereolithography apparatus (SLA).

3D Systems offers SLA 250, SLA 500, SLA 3500, SLA 5000, SLA 7000, Viper si^2, and so forth. The primary differences between these models are build chamber size, speed, cost, and types of lasers. D-MEC's solid creation system also uses UV laser to draw cross-sectional patterns on a resin solution surface. The rapid prototyping principle is very similar to that of the stereo-lithography process. D-MEC offers several models of prototyping solutions for companies. The primary differences between these models are types of lasers used, the build chamber size, and cost. CMET also offers several SOUP models that differ mainly in build size and cost. Because of its merger with Teijin Seiki, CMET also sells Soliform rapid prototyping systems, which were originally developed by Teijin Seiki. All of these are similar to 3D Systems stereolithography process.

PROBLEMS

3.1. Describe the principle of the stereolithography process.

3.2. Describe how the stereolithography apparatus (SLA) system works.

3.3. Describe the process flow for D-MEC's solid creation system (SCS).

3.4. Based on the specification of lasers for D-MEC's SCS, discuss which model you would choose to buy and why.

3.5. A company is using an SLA 250 machine for its prototyping needs. The nominal laser power is 20 mW and the central spot size is 0.30 mm. The scanning speed of the laser is 220 mm/sec. Calculate (a) the average exposure, (b) exposure time of the laser, (c) average energy density of the laser exposure, and (d) the number of photons/cm^2 hitting the resin surface.

REFERENCES

1. C. C. Kai and K. F. Leong, *Rapid Prototyping: Principles and Applications in Manufacturing,* Wiley, Singapore, 1997.

2. P. Jacobs, *Rapid Prototyping and Manufacturing & Fundamentals of Stereolith-ography,* SME Publications, Dearborn, MI, 1992.

3. P. K. Wright, *21st Century Manufacturing.* Prentice Hall, Upper Saddle River, NJ, 2001.

4. *www.d-mec.co.jp.*

5. *www.cmet.co.jp.*

4

SOLID-BASED RP SYSTEMS

This chapter describes rapid prototyping systems whose initial form of the material is solid. The solid-based RP systems that are studied include Stratasys' FDM system, Helysis' LOM system, Solidscape's 3D printing and deposition milling, 3D Systems' multijet modeling system, and KIRA's selective adhesive and hot pass (SAHP) system. For each system, a company profile will be provided in addition to information about the principle, the process, the products, the advantages and disadvantages, and applications.

4.1 STRATASYS' FUSED DEPOSITION MODELING

4.1.1 Company Profile

In 1988, Scott Crump, the president and CEO of Stratasys Inc,* developed the fused deposition modeling (FDM) process. Stratasys' mission was to provide design engineers with cost-effective, environmentally safe, rapid modeling and prototyping solutions. FDM generates three-dimensional models using an extrusion process. The first U.S. FDM patent was issued in 1992.

The FDM line of rapid prototyping systems allows designers to produce accurate, functional prototypes for testing and final design verification. These products can also be used to produce tooling patterns and masters for casting, room temperature vulcanized (RTV) molds and spray-metal tooling applica-

---

* Address: Stratasys, Inc., 14950 Martin Drive, Eden Prairie, MN 55344-2020. Toll-free number: 888-937-3010; telephone: +1.952.937.3000; fax: +1.952.937.0070; e-mail: info@stratasys.com; website: www.stratasys.com.

tions. Current FDM systems offered include the FDM Vantage, FDM Titan, FDM 8000, and FDM Maxum. Prodigy Plus models. In addition, Stratasys has the exclusive right in North America to effectively and profitably market Eden 260 and Eden 333 systems which are produced by Object company in Israel.

Using patented Fused Deposition Modeling (FDM) and PolyJet (available in the Eden 333 in North America only) rapid prototyping processes, Stratasys RP systems create precision three-dimensional prototyping parts directly from 3D CAD systems for use in testing forms, fit and function throughout the design and development process.

Stratasys systems allow design engineers to model highly complex geometries in a wide range of high-performance engineering materials—right from their workstation or network—with in-office RP systems that require no chemical post processing, special venting or facility modification.

Stratasys, Inc. is a worldwide provider of office prototyping and 3D printing solutions. The company manufactures rapid prototyping (RP) and 3D printing systems for the automotive, aerospace, industrial, recreational, electronic, medical, consumer products OEM, and education markets. The company's patented Fused Deposition Modeling (FDM) rapid prototyping processes create precision three-dimensional plastic prototyping parts directly from 3D computer-aided-design (CAD) systems. Stratasys holds more than 110 granted and pending patents worldwide focused on rapid prototyping. According to Wohlers Report 2004, Stratasys was the RP market unit leader in 2003 with 37 percent of all RP systems shipped worldwide that year.

The company has created a family of RP products with a wide range of modeling materials. The product line is designed to meet customers' needs across a spectrum of industries and throughout every phase of product development. The company is the largest RP retailer in the world.

4.1.2 Principles

Rather than being tooled from a solid body, 3D models made by rapid prototyping technology are built by the addition of materials and/or the phase transition of materials from a fluid to a liquid state [1].

While many different approaches exist in the methodology of rapid prototyping, this section focuses on fused deposition modeling. In this method, a model created in a CAD program is imported into a software program specifically designed to work with the FDM machine, in this case, Insight. The data for the model is saved as an STL (stereolithography) file for interpretation by Insight [2].

This program then uses a slicing algorithm to divide (or slice) the computer graphical solid model into layers (Fig. 4.1), each representing a layer of the material to be used [3]. The information for the model is then downloaded to the FDM machine as an SML file (Stratasys machine language) file. The machine builds the prototype by extruding a thin stream of semiliquid ABS

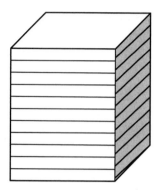

Figure 4.1   Sliced model.

plastic, referred to as a road, from a heated head or tip. The road is laid across the $X–Y$ plane one layer at a time, each layer corresponding to a slice determined by the Insight program. Each slice has the same height as a road [2].

To build each layer, the extrusion head deposits the outline of the layer first and then fills each layer according to the raster dimensions set by the operator. The raster fill of the outline is a series of straight-line segments that link together and are uniformly spaced (Fig. 4.2). As each layer is deposited, it fuses to the previous one, creating a solid model [2].

The ABS P400 plastic used is a durable ABS-based material that is appropriate for concept models as well as for testing of form, fit, and some function. The ABS P400 plastic is impact resistant, has a relatively high tensile strength, and is heat, scratch, and chemical resistant. This material has a high thermal expansion for a plastic and is lower in cost than most engineering thermoplastics. However, ABS plastic has limited weather resistance and is not very resistant to solvents [4]. Table 4.1 shows some of the advantages and disadvantages of ABS plastic.

In addition to ABS, Stratasys also uses the following materials:

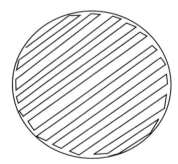

Figure 4.2   Contour and raster.

TABLE 4.1   Advantages and Disadvantages of ABS Plastic

Advantages	Disadvantages
Impact resistant	Not well-suited for direct
High tensile strength	tooling
Heat resistant	
Chemical resistant	
Scratch resistant	

- ABSi (sterilizable & translucent)
- Polycarvonate (PC)
- PC-ISO Medical grade PC
- Polyphenylsulfone (PPSF)
- PC-ABS blend

The Stratasys FMD 1650 machine has a heated head (Fig. 4.3) that extrudes the breakaway support system. As with the ABS plastic, this material is deposited as layers on the $X–Y$ plane and serves to support any overhanging portions of the prototype's geometry. As the name implies, the support material is easily broken away from the model. The Stratasys FDM 1650 was used for the case study described later in this chapter.

The Stratasys FDM 1650 is capable of building models with maximum dimensions of $9.4 \times 10 \times 10$ in., and has slice resolution (layer height)

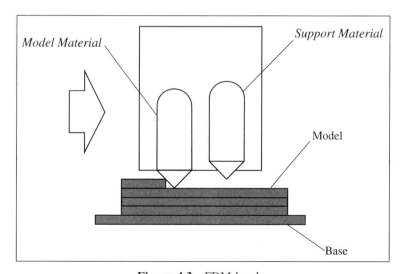

Figure 4.3   FDM heads.

ranging from 0.005 to 0.020 in. The width of the road varies according to the size of the tip being used or can be adjusted in the software. Although FDM 1650 has been a very reliable system, the company does not produce or support this system anymore.

4.1.3 Process

The FDM process forms three-dimensional objects from CAD-generated solid or surface models. A temperature-controlled head extrudes thermoplastic material layer by layer. The designed object emerges as a solid three-dimensional part without the need for tooling.

The process begins with the design of a geometric model on a CAD workstation. The design is imported into Stratasys' easy-to-use software, Insight, which mathematically slices the .stl file into horizontal layers. The SupportWorks software, also included, automatically generates supports if needed. The operator creates tool paths with the touch of a button. The system operates in the X, Y, and Z axes, in effect, drawing the model one layer at a time [5].

Thermoplastic modeling material, 0.070 in. (0.178 cm) in diameter, feeds into the temperature-controlled FDM extrusion head, where it is heated to a semiliquid state. The head extrudes and deposits the material in ultra-thin layers onto a fixtureless base. The head directs the material into place with precision, and, as each layer is extruded, it bonds to the previous layer and solidifies. A flow diagram of the FDM process is shown in Figure 4.4.

The FDM systems are modular—allowing customers to increase system capabilities as their prototyping needs increase in complexity. The basic system includes all the necessary components to generate models and prototypes

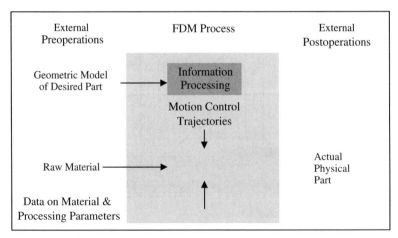

Figure 4.4 Flow diagram of FDM process.

in one material. Modules are available to increase capabilities with a variety of model-making materials.

4.1.4 Products

Over the years, Stratasys Inc. has developed many different products to facilitate different needs. Table 4.2 shows the technical specifications of several solid-based RP systems from Stratasys. A brief description of each of the products follows:

FDM Vantage SE   FDM Vintage is a versatile machine to create ideal system. Built on the proven T-class performance platform, one can configure this FDM Vantage system for enhanced speed, expanded build envelope or the high performance engineering material of one's choice (Fig. 4.5).

FDM Vantage T-Class System Highlights
- Excellent mid-range FDM solution with versatile system configuration
- Multiple high-performance engineering materials
- Powerful Insight™ software
- Office environment—requires no special venting
- Future upgrade options

FDM Titan   The Titan is the solution for building large, strong and durable parts. The Titan features multiple material supports and one of the largest build chambers in its class (Fig. 4.6). The Titan creates prototypes that have superior impact strength and also resist heat and corrosive agents such as oil, gasoline, and acids. Materials used in this system are ABS and polycarbonate. Applications for this system include:

- Designing a part at every phase of the development process
- Testing function to fit and form
- Simulating product performance
- Producing prototype tooling

FDM Maxum   The FDM Maxum is the fastest prototyping system offered by Stratasys, operating 50% faster than previous machines. The FDM Maxum includes WaterWorks-soluble support system and Insight preprocessing software. The FDM houses a high-capacity build chamber that accommodates very large parts or several smaller parts. The patented MagnaDrive system moves both FDM extrusion heads using magnetism and a thin cushion of air to achieve uniformly smooth, detailed, and accurate models (Figure 4.7). The material used by this system is ABS white. Applications for this system include:

TABLE 4.2 Solid-Based RP Systems by Stratasys

Machine	FDM Vantage SE	FDM Titan	FDM Maxum	FDM Prodigy
Work volume	16 × 14 × 16 in.	Maximum size 16 × 14 × 16 in.	Maximum size 23.6 × 19.7 × 23.6 in.	Maximum size 8 × 8 × 12 in.
Software interface	Insight software automatically imports STL files, slices the file and generates the FDM Vantage extrusion paths and necessary support structure.	Insight software imports STL files, automatically slices the file and generates the FDM Titan extrusion paths and necessary support structures and necessary support structures.	Insight software imports STL files, automatically slices the file and generates the FDM Maxum extrusion paths and necessary support structures	Insight software imports STL files, automatically slices the file and generates the Prodigy Plus extrusion paths and necessary support structures
Layer thickness	ABS: .010 in. (.254 mm). Polycarbonate: .007 in. or 0.10 in. (.178 mm or .254 mm). circuit required).	ABS: .005 in. (.127 mm) or .010 in. (.254 mm). Polycarbonate: .007 in. (.178 mm) or 0.010 in. (.254 mm). Polyphenylsulfone: .010 in. (.254 mm).	ABS: .005 in. (.127 mm), .007 in. (.178 mm) or .010 in. (.254 mm). ABSi: .007 in. (.178 mm) or .010 in. (.254 mm).	User Selectable .007 in. (.178 mm). .010 in. (.245 mm). .013 in. (.33 mm)
Accuracy precision	+/− .005 in.	+/− .005 in.	+/− .005 in.	+/− .005 in.
Material supply	ABS and Polycarbonate	ABS, Polycarbonate, and Polyphenylsulfone	ABS and ABSi plastic	ABS plastic in white (standard), blue, yellow, black, red or green. Custom colors available
Size	50.25 in. wide × 34.4 in. deep × 76.75 in. high	50.25 in. wide × 34.75 in. deep × 78 in. high	88 in. × 44 in. deep × 78 in. high	27 in. wide × 34 in. deep × 41 in. high
Weight	1600 lb	1600 lb	2500 lb	282 lb

Figure 4.5   Stratasys FDM Vantage RP Machine. (Courtesy of Stratasys, Inc.)

Figure 4.6   Stratasys FDM Titan RP machine. (Courtesy of Stratasys, Inc.)

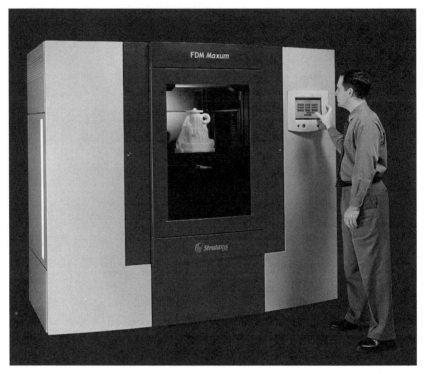

Figure 4.7   Stratasys FDM Maxum RP machine. (Courtesy of Stratasys, Inc.)

- Designing a part at every phase of the development process
- Testing function to fit and form
- Simulating product performance
- Producing prototype tooling
- Large ABS models

Prodigy Plus   The Prodigy Plus quickly transforms conceptual designs to functional prototypes and saves critical time spent on development. The Prodigy Plus uses the FDM process along with the WaterWorks support system in order to deliver a mechanism with moving parts built with no additional work needed. The Prodigy Plus uses the Insight processing software to quickly prepare files for manufacturing. It features a work volume of 8 × 8 × 12 in., produces layer thickness of 0.010 to 0.013 in., uses ABS white, blue, yellow, black, red, or green, and weighs a total of 300 lb. The Prodigy Plus system (Fig. 4.8) gives designers the ability to make durable models with moving parts that are built preassembled, test form, fit, and function, which are all within the networked office environment.

Figure 4.8 Stratasys FDM Prodigy Plus RP machine. (Courtesy of Stratasys, Inc.)

Dimension In February 2002, Stratasys introduced the Dimension product. The Dimension produces parts in ABS plastic in order for engineers to perform functional tests. The system uses the Break Away Support System (BASS) rather than the newer improved WaterWorks-soluble support system. Included with the Dimension system comes the Catalyst software providing queue management capabilities, build time, material status, and other machine status information. Dimension (Fig. 4.9) runs unattended and can provide system status and build completion information via e-mail, pager, or the Internet.

Figure 4.9 Stratasys dimension RP machine. (Courtesy of Stratasys, Inc.)

Eden 260*   Eden260 from Stratasys is compact, high resolution in-office rapid prototyping system (Fig. 4.10). Designed to fit smaller spaces and tight budgets, the Eden260 uses Polyjet technology to precisely jet UV plastics in super fine layers and high resolution. The technical specifications of Eden260 and Eden333 are shown in Table 4.3.

Some of the Eden260 system highlights are:

- Compact, space-saving system design
- Smooth surface/fine feature part output
- Fast build time
- Powerful Object Studio software
- Office operation with no special venting

Eden 333*   Eden333 from Stratasys is a smooth, clean, multiple material RP. It brings the convenience and control of fine feature RP to our network or desktop—with range of UV plastics. Polyjet technology precisely jets UV plastic in super fine layers down to 16 microns (Fig. 4.11).

Some of the Eden333 system highlights are:

- Smooth surface/fine feature part output
- Multiple UV plastic materials available

Figure 4.10   Stratasys Eden 260. (Courtesy of Stratasys, Inc.)

---

*These systems are manufactured by Object Geometries in Israel. Stratasys has the exclusive right to sell these systems in North America.

TABLE 4.3 Technical Specifications of Eden Systems K

Machine	Eden260	Eden333
Work volume	10.2 × 9.8 × 8.1 in.	13.4 × 13 × 7.9 in.
Software interface	Studio Software suggests the build orientation based on part size or required build speed, then automatically processes the STL files and generates necessary support structures in real time. The software creates a precise path for eight jetting heads to simultaneously deposit UV plastic for model building, and a second for gel-like UV plastic material for any necessary support structures.	Studio Software suggests the build orientation based on part size or required build speed, then automatically processes the STL files and generates necessary support structures in real time. The software creates a precise path for eight jetting heads to simultaneously deposit UV plastic for model building, and a second for gel-like UV plastic material for any necessary support structures.
Layer thickness	Horizontal build layers down to 16 microns	Horizontal build layers down to 16 microns
Accuracy precision	X axis: 600 dpi, Y axis: 300 dpi, Z axis: 1600 dpi	X axis: 600 dpi, Y axis: 300 dpi, Z axis: 1600 dpi
Material supply	Building and support materials are UV plastics cured by exposure to UV light	Building and support materials are UV plastics cured by exposure to UV light
Size	34.3 in. wide × 30 in. deep × 47.2 in. high	52 in. wide × 38 in. deep × 46 in. high
Weight	620 lbs	900 lbs

Figure 4.11   Stratasys Eden 333. (Courtesy of Stratasys, Inc.)

- Fast build times
- Powerful Object Studio software
- Office operation with no special venting
- Ideal complement to FDM RP systems

4.1.5   Advantages and Disadvantages

Advantages

- No postcuring
- Variety of materials
- Easy material changeover
- Low end, economical, office environment machines
- Fast build speeds on small/hollow geometries
- Low entry price

Disadvantages

- Surface finish is not as good as production parts
- Weak Z axis
- Slow build speed on larger/dense parts
- Some of the disadvantages mentioned here are very general in nature and may not apply to all the machines of Stratasys.

4.1.6   Applications and Uses

- Proof of concept
- Engineering review/red lines

- Marketing tools
- Vendor quotes
- Plating/vaccuumizing
- Investment casting patterns
- RTV models
- Sand-cast patterns
- Spray metal tools
- Product mockups
- Manufacturing requirements
- Masters for tooling
- Fit, form, and function

Concept Modeling   Concept models enable designers to visualize their designs three dimensionally. Designers use models for quick design verification and as communication tools. In contrast, traditional methods of concept modeling typically require about 1 to 4 weeks for development.

Concurrent Engineering   Concurrent engineering is a philosophy in which all departments work together on product development from the beginning. With this approach, errors can be caught very early in the development cycle, saving time and labor costs. Concurrent engineering also involves the use of a single database from design engineering and manufacturing to quality control. Stratasys' rapid prototyping and modeling systems facilitate the concurrent engineering process based on the strength of the database used to drive the system.

4.2   HELYSIS LAMINATED OBJECT MANUFACTURING SYSTEM

4.2.1   Company Profile

Founded in 1985, Helysis* is one of the early pioneers in rapid prototyping. Early research and development resulted in the creation of the laminated object manufacturing (LOM) process. Helysis has received a total of five patents for the LOM process. Although Helysis started off strong, the company has recently gone out of business. However, the LOM process is still widely used with many machines still on the market and supported by third-party companies.

---

*Address: Helysis, Inc., 1000 E. Dominguez Street, Carson, CA 90746-3608.

4.2.2 Principles

The LOM process generates a part by laser trimming materials to a desired cross section. The material is delivered in sheet form. The cross sections cut from the sheets are laminated into a solid block form using a thermal adhesive coating. The automatic process continues until all the cross sections of a part are created, then the final object is removed from the excess material [6].

4.2.3 Process

The LOM process (Fig. 4.12) uses a unique additive/subtractive process by which the adhesive-coated sheet material is bonded to the layer beneath and then laser cut to exact dimensions. The part does not go through a phase change. Building layer by layer, a three-dimensional product emerges that can be used for visualization, form, fit, function, or tooling. This unique additive process is especially advantageous for building a single large part or multiple smaller parts faster.

4.2.4 Products

The Helysis systems use the unique LOM process by which a thin adhesive-coated material is laser cut and bonded to the layer beneath. Building layer by layer, a three-dimensional product emerges. The top-of-the-line 2030H offers the largest build envelope and lowest total cost of ownership of any major rapid prototyping technology. The 1015Plus offers all of the advantages of the 2030H on a smaller platform. A comparison of the two systems is shown in Table 4.4.

4.2.5 Advantages and Disadvantages

Advantages

- No postcuring
- No support structure needed

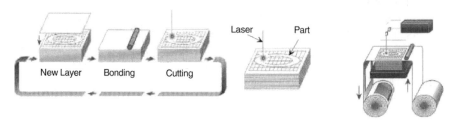

Figure 4.12   LOM process. (Courtesy of Helysis, Inc.)

TABLE 4.4 Specifications of LOM Models: Solidscape

	LOM-2030H	LOM-1015Plus
Build envelope	32 × 22 × 20 in. (81.3 × 55.9 × 50.8 cm)	15 × 10 × 14 in. (38.1 × 35.6 × 35.6 cm)
Laser beam diameter	From 0.008 to 0.010 in.	From 0.008 to 0.010 in.
Motion system repeatability	0.002 in. (0.0508 mm)	0.002 in. (0.0508 mm)
Material thickness	From 0.003 to 0.008 in.	From 0.003 to 0.008 in.
X-Y motion system speed	18 in./sec (457 mm/sec)	18 in./sec (457 mm/sec)

- Low maintenance/material cost
- No material phase change/warpage
- Can fabricate large parts

Disadvantages

- Surface finish appearance
- Machinability of parts is geometry dependent
- Parts do not perform well in humid or wet environment
- Excess material requires removal; internal cavities can be difficult
- Environmental concerns; outside venting is required

4.2.6 Applications and Uses

The LOM process is used by a wide variety of industries, from automotive to footwear. Some of the following industries use the LOM process:

- Aerospace
- Art
- Automotive
- Consumer products (toys, etc.)
- Footwear
- Science
- Sporting goods
- University research

Helysis' LOM system offers the speed, flexibility, large capacity, and low cost that are critical factors that manufacturers and designers are seeking to reduce in the product development cycle.

4.3   SOLIDSCAPE'S 3D PRINTING AND DEPOSITION MILLING

4.3.1   Company Profile

Solidscape,* Inc. (formerly known as Sanders Prototype, Inc.) designs, develops, manufactures, and sells rapid prototyping hardware and software. Solidscape's PatternMaster and ModelMaker are used in conjunction with CAD systems to automatically create physical models, industrial patterns, and prototypes directly from computer-generated designs. Solidscape's RP systems have been used in the design and development of products including medical/dental devices, aerospace components, automotive parts, sporting goods, and fine jewelry.

Solidscape is a privately owned company that incorporated Sanders Prototype, Inc. in February 1994. The original research and development for the ModelMaker technology was carried out between 1992 and 1994 under funding by SDI, a research company, which is principally owned and managed by Solidscape's founder, Rayden Sanders.

4.3.2   Principles of Solidscape Prototyping

Solidscape introduced the 3D Printing technology that uses dual ink printer heads to deposit both the thermoplastic part material and the wax support structure. The device uses a milling operation to create very thin, precise layers. The ModelMaker from Solidscape, Inc., produces parts by printing successive layers on foam substrate attached to an elevator table. Each layer consists of a horizontal cross section of the model. After a layer is completed, the table is lowered slightly, the layer is machined, and the next layer is printed on top of the last one. In this manner the thin cross sections are stacked on top of each other to build up the three-dimensional model [7].

Many models have an overhang at some point, that is, a layer that is larger than the one below it. This can create a problem since there is nothing to support the overhung portion when it is printed; instead, the overhung portion will fall down to the table surface below. To prevent this, temporary support material must be added to the lower layers wherever an overhang will occur. The support material is removed after the model is completed.

4.3.3   Process

The steps for making prototypes using the Solidscape 3D printing and deposition process are as follows:

---

*Address: Solidscape, Inc., 316 Daniel Webster Highway, Merrimack, NH 03054-4115. Telephone: (603) 429-9700; fax: (603) 424-1850; e-mail: precision@solid-scape.com.

Slicing the Part   A three-dimensional solid model of the part that is to be produced is created by either Pro/Engineer, AutoCAD, or any other comparable CAD software package. The model must then be saved in any of the following formats:

- STL (ASCII or binary)
- AutoCAD DXF
- 3D Systems' SLC
- 3D Systems' HPGL

These files can be read and loaded into the Model Works software provided with the RP machine. Once the model is loaded, the model is ready to be sliced into layers for the prototyping process. The user defines the slice configuration. Variables in the configuration include: slice thickness, printing pattern density, extra support material, extra cooling time, and jet check parameters. Once loaded, the part can be viewed as it is built. At this point the part is moved to the origin of the building surface. The part can also be rotated in any angle until the user is satisfied with the building orientation. The model is then sliced using the defined configuration and is ready for the prototyping process.

Building the Model   A piece of foam substrate is placed onto the build table by the operator. The substrate is then milled until a sufficiently smooth surface is attained. The machine is then allowed to proceed with the model-making process. This process is fully automated and will continue until the model is built. The finished model contains support material that must be removed. This is done by placing the model in a solution, which dissolves the support material and the attached base substrate.

4.3.4   Products

Solidscape's PatternMaster and ModelMaker prototyping systems combine a proprietary thermoplastic ink jetting technology and high-precision milling with the proprietary ModelWorks graphical front-end software to build prototype models or patterns that have an achievable accuracy of 0.001 in. (0.025 mm) per inch in the X, Y, and Z dimensions and a completed model surface finish of 32 to 63 μin. The PatternMaster and ModelMaker can build in slices, or layers, as small as 0.5 thousandths of an inch, which is an order of magnitude better than other rapid prototyping systems. These attributes make the PatternMaster and ModelMaker the rapid prototyping leaders in producing precision models and patterns for investment casting or plastic molding with little or no benchwork. With a slice thickness of 0.5 thousandths of an inch and the excellent surface finish, the need for lengthy benchwork associated with other RP systems is eliminated. Technical data for the PatternMaster and ModelMaker are given in Table 4.5.

While both the Model Maker II and Patternmaster are still in the market, the company does not support these systems anymore. The company has developed the following two systems:

T612 Benchtop   The T612 Benchtop system reduces costs and time-to-market a product by translating CAD designs into solid 3D models so accurate that one can go beyond concept modeling to produce tooling grade patterns ready for casting or mold-making. Now one can put the power of rapid prototyping to work for even one's most detailed, intricate, and complex parts. So now you customers can evaluate designs and processes—quickly and realistically—before committing to hard tooling or other major investments of time or money.

T66 Benchtop   The T66 Benchtop also reduces costs and time-to-market by translating CAD designs from most jewelry and engineering design software into 3D casting patterns. Grow Models of most intricate and complex designs can be grown with the assurance of results that are geometrically perfect and, in many cases, unable to be produced by hand crafting or by milling systems. Undercuts, overhangs and cavities are no problem because of Solidscape's dissolvable support material structure. The T66 Benchtop operates on standard current, uses non-toxic build materials and has a small footprint . . . making it truly office friendly. And the consumables are inexpensive to use when compared to other model building system technologies. The technical specifications of the two models are shown in Table 4.6.

4.3.5   Advantages and Disadvantages

Advantages

- Precision models with 0.0005- to 0.003-in. layer thickness
- Ultra-smooth surface finishes of 70 μin.

TABLE 4.5   3D Printing & Deposition Milling: Solidscape

	ModelMaker II	PatternMaster
Build envelope	12 × 6 × 8.5 in. (30.48 × 15.24 × 21.59 cm)	12 × 6 × 8.5 in. (30.48 × 15.24 × 21.59 cm)
Build layer	From 0.0005 (0.013) to 0.0030 in. (0.076 mm)	From 0.0005 (0.013) to 0.0030 in. (0.076 mm)
Achievable accuracy	± 0.001 in. (0.025 mm) per inch in X, Y, and Z dimensions	± 0.001 in. (0.025 mm) per inch across the X, Y, and Z dimensions
Surface finish	32–63 μin. (rms)	32–63 μin. (rms)
Minimum feature size	0.010 in. (0.254 mm)	0.010 in. (0.254 mm)

TABLE 4.6 Technical Specifications of T612 Benchtop and T66 Benchtop

	T612 Benchtop	T66 Benchtop

Build envelope and surface finish	• Ultra-high resolution: 25M addressable drops per square inch (25.4 mm^2)
	 • Build envelope: X = 12 in. (30.48 cm)
	   Y = 6 in. (15.24 cm)
	   Z = 6 in. (15.24 cm)
	 • Build layer: 0.0005 in. (0.013 mm) to 0.0030 in. (0.076 mm)
	 • Achievable accuracy: +/− 0.001 in. (0.025 mm) per inch in X, Y, and Z dimensions
	 • Surface finish: 32–63 microinches (RMS)
	 • Minimum feature size: 0.010 in. (0.245 mm)
		• Ultra-high resolution: 25M addressable drops per square inch (25.4 mm^2)
		 • Build envelope: 6 × 6 × 6 in. (15.24 × 15.24 × 15.24 cm)
		 • Build layer: 0.0005 in. (0.013 mm) to 0.0030 in. (0.076 mm)
		 • Achievable accuracy: +/− 0.001 in. (0.025 mm) per inch in X, Y, and Z dimensions
		 • Surface finish: 32–63 microinches (RMS)
		 • Minimum features size: 0.010 in. (0.254 mm)

Materials	• Works with silicone, RTV, epoxy, and other elastomeric molds • Can be used to build injection molding tools	• Works with silicone, RTV, epoxy, and other elastomeric molds • Can be used to build injection molding tools
Dimensions and operate system	• Small Footprint: 28.0 in. (71.12 cm) width × 19.5 in. 49.53 cm) depth × 19.5 in. (49.53 cm) height • Works with standard CAD files: .STL, .SLC • Operates with standard IBM-compatible PC, running Microsoft Windows 2000 Professional, XP Professional or ME	• Small Footprint: 22.0 in. (55.88 cm) wide × 19.5 in. (49.53 cm) deep × 19.5 in. (49.53 cm) high • Works with standard CAD files: .STL. .SLC • Operates with standard IBM-compatible PC, running Microsoft Windows 2000 Professional, XP Professional or ME

- Castability of all object patterns
- Hands-free postprocessing

Disadvantage

- Thin layers require longer build times

4.3.6   Applications and Uses

PatternMaster and ModelMaker systems are marketed worldwide and designed for numerous manufacturing applications including jewelry, jet engine turbine blades, medical instruments and prosthetics, consumer goods, automotive parts, electronics, and many other high-precision products.

4.4   3D SYSTEMS' MULTIJET MODELING SYSTEM

4.4.1   Company Profile

3D Systems was founded in 1986 by Charles Hull and Raymond Freed. It pioneered commercial RP technology with the stereolithography apparatus, or SLA as it is commonly called. In 1996, 3D Systems introduced its multijet modeling (MJM) technology under the product name Actua. The company's details are found in Chapter 3.

4.4.2   Principles

The principle underlying the Actua 2100 is the layering principle used in other rapid prototyping and manufacturing (RP&M) systems. MJM builds models using a technique akin to ink-jet or phase-change printing, applied in three dimensions. A "print" head comprised of 96 jets oriented in a linear array builds models in successive layers, with each jet applying a special thermopolymer material only where required. If the part is wider than the MJM head, the platform repositions itself to continue building the layer.

4.4.3   Process

The process of the Actua 2100 is simple and fully automated. In the build process, the MJM head is positioned above the platform to start building the concept model from the base up. The head begins building the first layer by depositing material as it moves in the X direction. If the width of the part is wider than the MJM head, the platform is repositioned (Y axis) to continue building the layer in the X direction until the entire layer is completed. After one layer is complete, the platform is lowered and the building of the next layer begins in the same manner. The process continues with the repetition of build steps until the part is complete. After completion, the part is ready for instant removal and review with no further need for postcuring or postprocessing [8].

4.4.4   Products

The Actua 2100 (Fig. 4.13) from 3D Systems enables a designer to produce a three-dimensional model easily. It uses the MJM technology to decrease the time required to complete a model. Table 4.7 contains the specifications for the Actua 2100.

4.4.5   Advantages and Disadvantages

Advantages

- Fast, efficient, and economical
- Inexpensive thermoplastic material
- Geometry of model does not effect work time
- No need for special facilities; can be used directly in an office environment
- Networking capabilities allow several workstations to be connected to the machine

Disadvantages

- Requires postcuring
- Limited materials
- Low accuracy relative to other RP methods

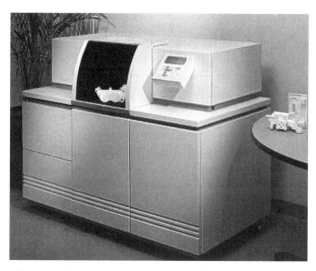

Figure 4.13   3D Systems' Actua 2100 multijet modeler.

TABLE 4.7   Specifications for Actua 2100

Model	Actua 2100
Technology	Multijet modeling (MJM)
Resolution	300 × 400 × 600 DPI (XYZ)
Maximum model size	250 × 190 × 200 mm (10 × 7.5 × 8 in.) (XYZ)
Build material	ThermoJet 2000 and ThermoJet 88 thermoplastic build material

4.4.6   Applications and Uses

The main application for the MJM is producing concept models for visualization and proofing during the early design process. It is meant to function in an office environment within the immediate vicinity of CAD workstations.

4.5   KIRA'S SELECTIVE ADHESIVE AND HOT PASS (SAHP) SYSTEM

4.5.1   Company Profile

KIRA,* a Japanese company, has worked aggressively since 1944 to gain a leading position as a machine tool builder. KIRA's standard machines and special-purpose machine tools have been widely accepted by industry, including leading manufacturers. The technology developed by KIRA's varied manufacturing experiences has provided the foundation to engineer and manufacture CNC machines and CNC special-purpose machine tools.

4.5.2   Principles

This is an RP system that fabricates shapes by laminating and cutting layers of paper in accordance with 3D CAD data. Paper lamination technology (PLT) has been under research and development for over 10 years. PLT has recently achieved a level of economy, speed, and reliability that ranks it as one of the best RP systems for reducing development time and cost.

4.5.3   Process

The resin powder is applied on a sheet of paper using a typical laser stream printer and is referred to as the xerography process. A sheet alignment mechanism adjusts the sheet of paper on to the previous layer of the model. Then, a hot press moves over them with the printed sheet pressed to a hot plate

*Address: Tomiyoshi shinden, Hazu-gun, Kira-cho, Aichi Pref., Japan. Telephone: 81-563-32-1161; fax: 81-563-32-3241; e-mail: info@kiracorp.co.jp; website: www.kiracorp.co.jp.

with high pressure. The temperature-controlled hot press melts the toner, which adheres the sheets together. The hot press also flattens the top surface and prevents bubbles between sheets. Then, based on plotting data, a mechanical cutter cuts the top layer of the block along the contour of the section. These steps are repeated until the entire model is built, as seen in Figure 4.14. The excess material is removed and the surface may then be finished by mechanical means [9]. The various steps of the PLT process are shown below.

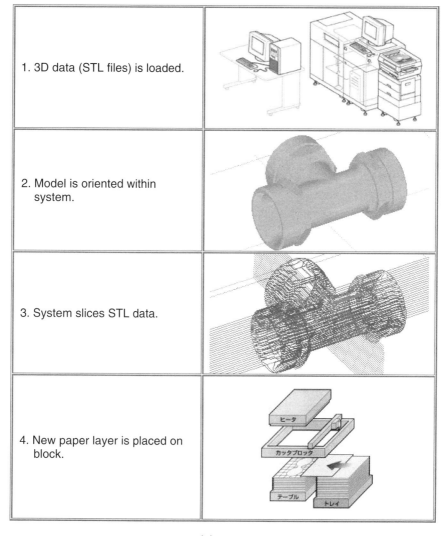

1. 3D data (STL files) is loaded.	
2. Model is oriented within system.	
3. System slices STL data.	
4. New paper layer is placed on block.	

(*a*)

Figure 4.14   Model-building process.

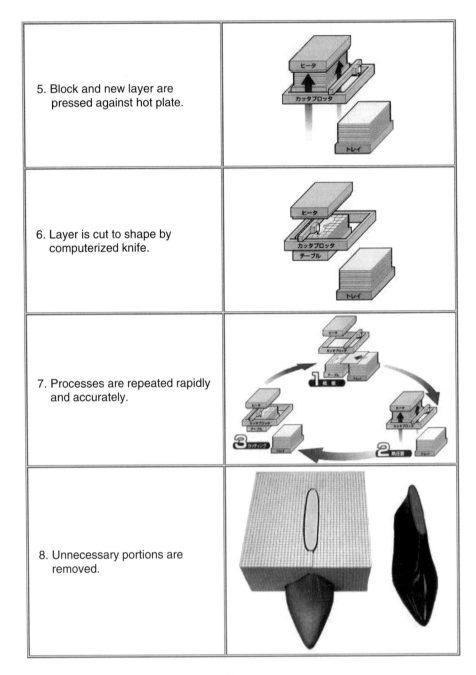

5. Block and new layer are pressed against hot plate.

6. Layer is cut to shape by computerized knife.

7. Processes are repeated rapidly and accurately.

8. Unnecessary portions are removed.

(b)

Figure 4.14 (*Continued*)

TABLE 4.8 Specifications of KIRA Corp. Products

System Name	KSC-50N	PLT-A4
Material	Plain paper and toner	Special paper
Maximum work size (mm)	400 × 280 × 300	190 × 280 × 200
Precision (mm)	XY plus minus 0.05,	XY plus minus 0.05,
	Z plus minus 0.1	Z plus minus 0.1

4.5.4 Products

KIRA Corporation Ltd. produces the KIRA Solid Center KSC-50 and the PLT-A4. The specifications for each are summarized in Table 4.8.

4.5.5 Advantages and Disadvantages

Advantages

- Fully automated process
- Model material is plain paper
- Hardness, flatness, and smooth surface finish
- Ease of installation
- Ease of finish

Disadvantages

- Needs postprocessing
- Needs supports

4.5.6 Applications and Uses

The main application areas are in conceptual modeling and visualization. KIRA's products have been used in the following industries:

- Automobile
- Electronics
- Education
- Camera
- Office automation

4.6 SUMMARY

In this chapter, the principles, operation, and applications of solid-based rapid prototyping systems were introduced and discussed. We have described the most proven solid-based RP technologies such as Stratasys' fused deposition modeling system, Helysis' laminated object manufacturing system, Solid-scapes's 3D printing and deposition milling, 3D System's multijet modeling system, and KIRA's selective adhesive and hot pass (SAHP) system.

Stratasys offers many different models to produce accurate and functional parts for visualization, testing, and design verification. Some of the FDM products marketed by Stratasys are: FDM 2000, FDM 3000, FDM Vantage, FDM Titan, FDM 8000, and FDM Maxum. The primary differences between these models are build size, speed, material type, and cost. In addition, Stratasys also sells Prodigy Plus and Dimension models, which are inexpensive concept modelers. The accuracy and strength of the prototypes made by these machines are not as good as those of other FDM machines, but they are marketed for design concept modeling in an office environment.

The LOM system uses a unique additive and subtractive process to bond adhesive-coated sheet material to make prototypes. This process is especially advantageous for large prototypes. Although Helysis is no longer in business, many of its products are still in use today.

Solidscape offers two types of machines, ModelMaker II and Pattern-Master. The differences in the two types of machines can be seen in the build envelope, accuracy, and surface finish of the products. The accuracy of these machines is about 10 times better than Stratasys' FDM-1650 and 3D Systems' SLA-250 machines. However, Solidscape's machines are much slower and the materials are much weaker.

3D System's MJM Actua Thermojet 2100 are concept modeling machines. Both machines produce wax prototypes with an array of inkjets. These machines are inexpensive, safe, and environmentally friendly. They compare favorably to Stratasys' Prodigy and Dimension machines.

KIRA offers a limited number of products that are mainly used for concept modeling and visualization. They are primarily sold in the Asian market.

PROBLEMS

4.1. Describe Stratasys' fused deposition modeling (FDM) process.

4.2. Describe the process flow of Helysis' laminated object manufacturing (LOM).

4.3. What are the differences between FDM and LOM? List the advantages and disadvantages of the two systems.

4.4. Describe the operating principle of 3D printing by Solidscape. How is 3D printing different from FDM? What are the strengths and weaknesses of 3D printing from Solidscape?

4.5. What are the types of rapid prototyping materials used in the FDM process? What is the difference between ABS and medical-grade ABS? Compare their mechanical properties in tabular form.

4.6. Using the machine specifications of the solid-based RP systems described in this chapter, which process is suitable for making the largest part? Which one will have the best accuracy? Which process is the slowest?

4.7. A mechanical engineering student has been asked to design a new housing for a computer. The finished part is roughly $15 \times 15 \times 10$ in. The part is essentially a thin shell with a $\frac{1}{8}$-in. thickness. Which process and machine would you suggest to produce the part and why?

4.8. A student of an art department has been asked to design and manufacture an earring with an intricate design. Describe your design, prototyping, and manufacturing process. Which machine is better for the fabrication of jewelry?

REFERENCES

1. R. Noorani, P. Gerencher, B. Fritz, M. Mendelson, O. Es-Said, and S. Dorman. *Impact of Rapid Prototyping on Product Development, Proceedings of the International Conference on Manufacturing Automation,* Hong Kong, pp. 884–889, 1997.

2. S. Crump. The Extrusion of Fused Deposition Modeling, Proceedings 3rd International Conference Rapid Prototyping, pp. 91–100, 1992.

3. C. C. Kai, and K. F. Leong. *Rapid Prototyping: Principles and Applications in Manufacturing,* Wiley, Singapore, 1997.

4. Stratasys, *FDM System Documentation,* F1650-5, F1650-6, F1650-11, G-8, G-9, and Q79, Stratasys, Eden Prairie, MN, 1996.

5. Stratasys, Inc. *www.stratasys.com,* 2002.

6. Helysis Inc., *http://www.helysis.com,* 2002.

7. Sanders Prototype, Inc. (Solidscape), *Model Maker Reference Manual,* Pine Valley Mill, Wilton, NH, 1996.

8. K. G. Cooper. *Rapid Prototyping Technology: Selection and Application.* Marcel Decker, New York, 2001.

9. Kira Corporation, *http://www.kiracorp.co.jp,* 2002.

5

POWDER-BASED RP SYSTEMS

This chapter describes the special group of solid-based rapid prototyping systems that primarily use powder as the basic medium for prototyping. Some of the systems in this group, such as selective laser sintering, bear similarities to liquid-based rapid prototyping systems. They generally use lasers to draw the part layer by layer; however, the medium used for building the model is a powder instead of photocurable resin. Other systems such as three-dimensional printing and multiphase jet solidification have similarities with the solid-based rapid prototyping systems. The common feature among the systems described in this chapter is that the material used for building the part or prototype is always a powder of some kind.

5.1 3D SYSTEMS' SELECTIVE LASER SINTERING PROCESS

5.1.1 Company Profile

Selective laser sintering (SLS) was developed and patented by the University of Texas at Austin and was subsequently commercialized by DTM Corporation. Established in 1987, DTM Corporation shipped its first commercial RP machine in 1992. With the support of B.F. Goodrich Company, DTM was successful in promoting the SLS process in its early days. In August 2001, 3D Systems bought DTM Corporation and further enhanced its position as a world leader.

5.1.2 Principles of SLS Process

The SLS process is based on two principles. First, parts are built by sintering when a laser beam hits a thin layer of powder material. The interaction of the laser with the powder raises the temperature to the softening point and results in particle bonding. The particles fuse themselves to the previous layer to form a solid. Second, the part is built layer by layer. Each layer of the building process contains the cross sections of one or many parts. The next layer of powder is deposited by a roller mechanism on top of the previously formed layer. That powder layer is then sintered to the previous layer by the laser. The SLS process is shown in Figure 5.1 [1].

Before discussing how the particles bond during the sintering process, it is necessary to investigate the packing of particles during sintering. The density of the gathered particles will have a significant effect on the bonding and the mechanical properties of the model. Generally, with higher packing density of the powder, better mechanical properties are expected. However, exposure parameters and scan pattern also have an effect on the mechanical properties of a part.

5.1.3   Principles of Sinter Bonding

Particles in successive layers are fused together to the previous layers by raising the temperature with the laser beam to above the glass transition temperature. The glass transition temperature is the temperature at which the model material softens. This often occurs just prior to the melting temperature at which the model material liquefies. As a result, the particles soften and cause the surfaces in contact with the particles to fuse together at the contact surface. Sintering differs from melting or fusing in that it joins the powder particle together without the deformations caused by the flow of molten material. After cooling, the powder particles form a matrix that has approximately the density of the particle material. The rapid prototyped part has a similar crystalline structure to a part prototyped using conventional methods.

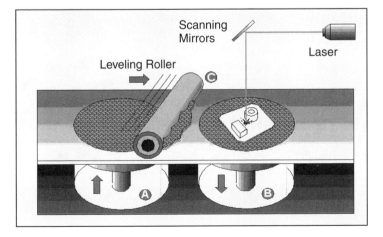

Figure 5.1   Selective laser sintering process. (Courtesy of 3D Systems, Inc.)

The crystalline structure presented in Figure 5.2 shows the fine-grain structure associated with the strength of the material [1].

The sintering process requires larger amounts of energy to bring the temperature of the particles to the transition temperature. The high power requirement can be reduced by heating the work volume to a temperature just below the sintering temperature of the part. Despite the energy advantages, there is a need for an inert gas environment in order to prevent oxidation or explosion of the powder. In addition, cooling of the chamber gas is also necessary.

The parameters that affect the performance and functionalities are the properties of the powdered material and its mechanical properties after sintering, the accuracy of the laser, the scanning pattern, the exposure parameters, and the resolution of the machine.

5.1.4 Process

The following steps describe the selective laser sintering process [1]:

1. An STL file is created from 3D CAD data.
2. The data is entered into the Vanguard SLS system in order to "slice" the solid model into successive cross sections to be traced by the laser.
3. A precision roller mechanism automatically spreads a thin layer of powdered SLS material across the build platform.
4. A cross section of the CAD file is sintered. Using data from the STL file, a CO_2 laser selectively draws a cross section of the object on the

Figure 5.2 Crystalline structure of rapid prototyped metal.

layer of powder. As the laser draws the cross section, it selectively "sinters" (heats and fuses) the powder creating a solid mass that represents one cross section of the part.

5. The system spreads and sinters layer after layer until the object is complete. Figure 5.3 shows a diagram of the SLS machine completing the sinter process for one layer.

6. Once the part is complete, it is removed from the part build chamber and any loose powder is removed.

7. The part can be finished as desired. The part is either used as is—or it can be sanded, annealed, coated, or painted depending on its intended application.

5.1.5 Laser Sintering ("LS") Materials

The SLS process and the SLS system are not limited to a specific class of materials. In theory, a wide range of thermoplastics, composites, ceramics, and metals can be used in the process. The main types of materials used in the SLS system are nontoxic, easy to use, store, and recycle. The material processing capabilities of SLS are as follows [1]:

1. CastForm material is similar to standard investment casting wax used by foundries but in powder form. This material is typically used to create wax patterns, which are then used in the investment casting process to create metal prototypes, cast tooling, and limited run parts. The

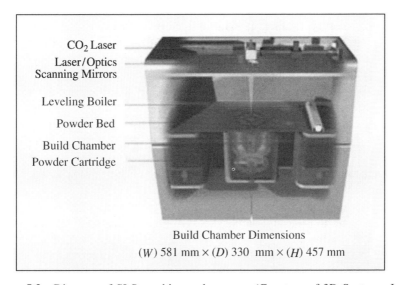

Figure 5.3 Diagram of SLS machine and process. (Courtesy of 3D Systems, Inc.)

CastForm wax leaves lower amounts of residue than most other casting waxes and therefore can be used to cast a wider variety of metal alloys. Rapid prototyped cast patterns can be used to augment the investment casting process development.

2. DuraForm material is a polycarbonate, industry standard, engineering thermoplastic. DuraForm is available in two types: the DuraForm PA material, which has the typical mechanical properties of thermoplastics, and the DuraForm GF material, which is 300% stronger and stiffer than DuraForm PA material. They are both suitable for creating concept and functional models and prototypes, durable investment casting patterns for metal prototypes, masters for the duplication process, and sand casting patterns. They are very durable and heat resistant, compatible with the investment casting process, can be built quickly, and are excellent for prototypes and patterns with fine features and thin parts.

3. Nylon: Nylon material is suitable for creating models and prototypes that can withstand and perform in a demanding environment. It is one of the most durable rapid prototyping materials currently available in the industry, and it offers substantial heat resistance and detail. Fine nylon is a variation of nylon and used to create fine-featured parts for working prototypes. It is durable, resistant to heat and chemicals, and is excellent when fine detail is required.

4. Somos 201 is a rubberlike material that possesses similar mechanical properties to the thermoplastic elastomers. This material is primarily used when part flexibility is desired.

5. LaserForm ST-100 is a 60% 420 stainless steel and 40% bronze matrix material with a thermoplastic binder. It is used in the infiltration process with the SLS systems platform to create mold cavity and core inserts for prototype tooling. It exhibits material properties that exceed those of 7075 aluminum, and, when used for metal tools, it is strong enough to produce more than 50,000 parts.

A comparison of the mechanical properties of 3D Systems' LS materials is presented in Table 5.1 [1].

5.1.6 Products

The Sinterstation HiQ System is the most recent 3D system selective laser sintering product. These SLS systems build small volumes of accurate, complex parts more rapidly and efficiently than other manufacturing processes. For one to several thousand parts, SLS system users can make plastic or metal parts directly from their CAD data, without ever making tools. For quantities ranging from 1000 to over 100,000, SLS technology also offers a rapid tooling solution. The machine specifications for the system are presented in Table 5.2. The Sinterstation HiQ system is presented in Figure 5.4 [1].

TABLE 5.1 Mechanical Properties of 3D Systems SLS Materials

Property	Method	Units	DuraForm	DuraForm GF	Somos 201
Tensile strength	ASTM D638	ksi (MPa)	6.4 (44)	5.5 (3.81)	2.24 (15.5)
Tensile modulus	ASTM D638	ksi (MPa)	232 (1600)	857 (5910)	
Young's Modulus	ASTM D638	ksi (MPa)			110
Elongation at break	ASTM D638	%	9.0	2.0	
Flexural strength	ASTM D790	ksi (MPa)			1.9 (13.4)
Flexural modulus	ASTM D790	ksi (MPa)	186 (1285)	479 (3300)	
Impact strength Notched Izod	ASTM D256	ft·lb/in. (J/m)	4.0 (214)	1.8 (96)	
Heat; deflection temperature	ASTM D648	°F (°C)	351	347	
@ 66 psi			(177)	(175)	
@ 264 psi			187 (86)	230 (110)	
Coefficient of thermal Expansion thermal conductivity	TMA [T<Tg]	ppm/°C BTU/ft·lb·°F (W/m °K)			
Glass transition, T_g	DMA	°F (°C)			
Dielectric constant −22°C, 50% RV, 5V, 1000 Hz		v/ft (v/mm)	2.9	3.7	2.9
Dielectric strength −22°C, 50% RV, in air, 5 V		v/ft (v/mm)	4.9×10^6 (1.6×10^4)	4.6×10^6 (1.5×10^4)	1.3×10^5 (4.1×10^3)
Hardness, Shore D	D2240			81	74 (A scale)
Density		g/cm³	0.97	1.40	0.91

TABLE 5.2 SLS Systems

Build chamber dimensions	$W14.5 \times D12.5 \times H17.5$ (in.)
Laser	25 or 100 W CO_2 (material dependant)
Beam delivery systems	Standard BDS max scan speed = 7500 mm/sec
Computer system	933 MHz Pentium III
Software	3D Systems software
CAD data	STL

5.1.7 Advantages and Disadvantages

Advantages

- *Speed* Move from CAD file to functional parts and tools in as little as one day, without programming, and with less highly skilled labor.
- *Plastic or Metal Parts Without Tools* Ideal for small volumes of parts that are produced much faster, and at less cost, than via tooling processes.
- *Complex Tooling Inserts* Ready for finishing in 2 days. The LaserForm material provides durable metal inserts without programming, milling, or machining. The cost of creating tooling is avoided.
- *Small Features, Complex Parts* Able to create accurate, complex parts and tools with fine features as thin as 0.5 to 0.6 mm (0.020 to 0.025 in.), without the need for programming, machining, or other time-consuming tooling processes.
- *Nest, Stack, and Add Parts During a Build* Provides more efficient productivity and quick response to keep up with consumer's demands.

Figure 5.4 Vanguard HiQ selective laser sintering machine. (Courtesy of 3D Systems, Inc.)

- *Unattended Operation*   Most parts require only 1 to 3 hr of the operator's time before final finishing begins.
- *Easy and Accurate CAD File Translation*   An STL file can be exported from a native file in one operation.
- *Reduced Cost per Part*   The speed of SLS systems allows more parts in less time and a lower cost per part.
- Build chamber dimensions (*W*) 381 mm × (*D*) 330 mm × (*H*) 457 mm.

Disadvantages

- *Initial Cost of System*   The initial cost of the SLS systems range from $250,000 to $500,000 depending on the options and peripherals acquired and excluding facility modifications.
- *Peripherals and Facility Requirements*   There are various peripherals necessary for optimum operation of the SLS systems, including a BOS table, air handler and shifter, and glass-bead blaster for finishing. Combined, these components take up about 30 ft^2 of floor space. In addition to a furnace for firing and infiltration of the metal parts is necessary for the rapid-tooling module, which requires special facility and safety requirements for operating gases and general maintenance. All of the systems have various facility requirements in addition to the hard-wired 240V/70A power requirements of the SLS itself, which also requires a large amount of floor space on the order of 200 ft^2. Finally, the SLS system weight combined is 6275 lb, which will require a sturdy floor to accommodate it.
- *Maintenance and Operation Costs*   Since the SLS systems are large and complex systems, the maintenance contracts currently run approximately $35,000 annually. Also the powders must be properly stored and recycled for further use. The power consumption of the system and all its peripherals can be high and must be taken into account, along with the smaller costs of expendable inert gases, build materials, and part finishing supplies.

5.1.8   Applications and Uses

The SLS process is used by many companies for a wide variety of applications. In this section, the special applications of the SLS process by two companies, BASTECH and SportRack, are mentioned.

BASTECH, Inc. recently put copper polyamide (PA) material through its paces and discovered a new, faster route to creating injection mold tooling that can produce as few as 20 or as many as several hundred parts.

One of the company's clients, an automotive OEM (original equipment manufacturer), needed 20 to 50 sets of brake reservoir parts produced in the intended production material, polypropylene. The parts had to withstand prolonged contact with brake fluid and perform like production parts during functional testing on a prototype vehicle. The parts are presented in Figure 5.5.

The brake reservoir consisted of two pieces that eventually would be welded together. Two tools were required to make these parts. One was made up of 13 pieces; the other 5 pieces. The final part, after welding, would measure approximately 110 by 154 by 41 mm.

Eight pieces of the 13-piece tool and four pieces of the 5-piece tool were made with copper PA on a Sinterstation system. The remaining pieces, very simple inserts and sleeves, were machined out of metal.

When the tools were complete, BASTECH readied them for injection molding. BASTECH ran 75 sets of parts on the copper polyamide tools, noticing very little wear. Company representatives note that the molds were quite sturdy and could easily have produced 100 to 150 sets of parts. Additionally, the SLS copper polyamide tooling withstood injection molding pressures of 400 psi and injection molding temperatures of 450°F [2].

BASTECH was pleased with these results, as was the customer. It also reports that a copper PA tool generated on a SLS system was about 25% of the cost and took only one half the time of a traditional steel or machined production tool.

SportRack, a company that produces automotive accessory prototypes for the world's top auto manufacturers, benefits from its in-house SLS system in many ways. One benefit is a significant reduction in prototype lead time. SportRack can now have prototype parts ready for its customers in a matter of hours or days, instead of weeks or months. The SLS system also gives SportRack the luxury of material versatility. To date, the company has created prototypes using DuraForm PA material, DuraForm GF material, and Somos 201 material.

SportRack reports that SLS system generated parts are durable. Many have been installed on test vehicles and challenged in a variety of ways, from

Figure 5.5 BASETECH prototyped brake reservoir.

grueling road tests to intense wind tunnel testing. The company recently tested a roof rack prototype produced with DuraForm material using the European Luggage Rack Test Standard or DIN test. The part exceeded DIN Level I standards. The roof rack is shown in Figure 5.6.

SportRack has also prototyped a complex ski rack carrier using DuraForm PA material. Initial prototype testing revealed the need for minor design adjustments, which SportRack made quickly. The second prototype performed so positively, SportRack moved immediately into production. It completely omitted the prototype injection molding phase, saving $69,000 in hard costs and 12 weeks of development time [2].

SportRack uses DuraForm PA material for parts with snap fits and DuraForm GF material for parts demanding increased stiffness and mechanical integrity. Both materials offer good dimensional stability and can be used to create prototypes and pattern masters for processes such as RTV tooling. SportRack also uses Somos materials for gaskets and seals and other parts that require pliability and flexibility.

Access to a faster, more efficient prototyping method has helped SportRack enjoy significant reductions in product development costs and a dramatic decrease in its dependency on outside prototyping vendors. In addition, SportRack can quickly create several different LS versions of a design without the costs and turnaround times associated with hard tooling.

5.2 MIT'S THREE-DIMENSIONAL PRINTING

5.2.1 Institution

The Three-Dimensional Printing (3DP) process is a manufacturing technology that has been invented and patented by the Massachusetts Institute of Technology (MIT) for the rapid and flexible production of prototypes, parts, and tooling directly from a CAD model. It is an extremely flexible system, capable of creating parts of any geometry and using any material including ceramics,

Figure 5.6 SportRack prototyped roof rack.

metals, polymers, and composites. This process is also capable of having local control over the material composition, microstructure, and surface texture [3].

MIT's 3DP research is funded by a noncompetitive industrial consortium and many national foundations such as the National Science Foundation (NSF), the Office of Naval Research (ONR), and the Defense Advanced Research Project Agency (DARPA). Consortium members exchange information and experience among each other and receive research and development information from the university–industry partnership.

5.2.2 Process

The 3DP process uses ink-jet printing technology to build parts in layers (Fig. 5.7). A slicing algorithm draws detailed information of the layer from the CAD model of the part. Each layer begins with a thin distribution of powdered material over the surface of a powder bed. Using the ink-jet printing technology, a binder material selectively joins the particles where the object is formed. A piston that supports the powder bed and the part-in-program, lowers the platform, deposits a fresh layer of powder, and binds the powder again. This layer-by-layer process continues until the part is completed. Following a heat treatment, unbound powder is removed, leaving the prototyped part.

During the build process, loose powder surrounding the bonded area serves as the support material, which also permits the part to have overhangs, undercuts, and complex internal geometry without the need for rigid support structures. The 3DP process is a very powerful prototyping process useful for the creation of functional parts and tooling directly from a CAD model. This technology pioneered the fabrication of ceramic parts and the direct fabrication of ceramic molds for casting.

MIT has licensed the 3DP technology to several companies in diverse fields. These companies are using MIT's 3DP process for specified fields of uses, although new applications are being developed every day. For example, MIT has issued Therics Inc. a license for the production of a time-release drug-delivery device. Using a multiple jet print head, extremely accurate

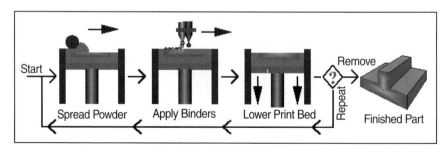

Figure 5.7   The 3DP process.

quantities of several drugs can be made into a biocompatible, water-soluble substrate designed to time-release these drugs into the bloodstream. Table 5.3 displays different companies licensed by MIT.

5.2.3 Major Applications

The following are examples of major applications of the 3DP process [3]:

1. *CAD Casting Metal Parts*   CAD casting represents a casting process where a mold is fabricated directly from a computer model without going through any intermediate steps. In this application, a ceramic shell with integral cores are fabricated directly from a CAD model, which results in streamlining the casting process as compared to traditional investment casting processes.

2. *Direct Metal Parts*   Metal parts using stainless steel, tungsten, and tungsten carbide can be created from metal powder using the 3DP process. Printed parts are postprocessed using injection molding techniques. The 3DP process has the advantage of adapting to a variety of material systems that allow the production of metallic/ceramic parts with novel compositions.

3. *Structural Ceramics*   The 3DP process has been used to prepare dense alumina components by spreading submicron alumina powders and printing a latex binder. The parts are then isostatically processed and sintered to increase the density of the component. The polymer binder is then removed by thermal decomposition.

4. *Functionally Gradient Material*   One of the strengths of the 3DP process is its ability to create parts with composite materials. For example,

TABLE 5.3   Company Licensing

License	Description
Z Corp.	Exclusive license to use 3DP for concept modeling and applications.
Soligen	Exclusive 3DP license for producing direct ceramic mold for metal casting. The process is known as direct shell production casting (DSPC).
Specific Surfaces	Holds 3DP license for filter manufacturing and uses Soligen's machine for production.
Extrude Hone	Extrude Hone's ProMetal Division sells the R4 and R10 rapid metal production systems. Both create solid metal objects from steel powder.
Therics	Therics uses its 3DP license to manufacture medical products including time-release medications.

a ceramic mold created using the 3DP process can be filled with particulate matter and then pressure infiltrated with a molten material.

5.2.4   Research and Development

The 3DP team at MIT, consisting of the researchers from both mechanical and material science departments, has been concentrating on improving the 3DP process for better manufacturing applications. The following are some of the areas of research and development on which the MIT team is focusing:

1. *Local Composition Control (LCC)*   The purpose of this program is to have the ability to create parts that have composition variations within them. Such local composition control has the potential to create new classes of components.

2. *Fabrication of Gradient Index Lenses (GRIN)*   The purpose of this research is to use 3DP to fabricate lenses that refract light due to the variations of index of refraction, instead of the geometry of the lens surface. Such lenses are called gradient index lenses, or GRIN lenses. GRIN lenses offer advantages in size and integration.

3. *Large, Hollow Metal Parts*   The purpose of this research is to enable the use of 3DP for the direct fabrication of large (>0.5 m) metal parts. The main emphasis is on the structural parts that take advantage of the ability to create complex internal geometry, such as truss structures, ribs, and so forth. The need for fabrication of structural metal components in small and moderate quantities will greatly extend the range of application of RP.

4. *Tooling for Lost-Foam Casting and Related Scale-up of 3DP*   The research is being used for the development of the ink-jet print head technology that will allow for a substantial scale-up of 3DP to a powder bed of 1×0.5 m in printing surface. MIT is also contributing to the development of material systems, which will enable larger parts to be cast.

5. *Tungsten Carbide Cutting Tools by 3DP*   This research is used to apply the 3DP process to the fabrication of tungsten carbide/cobalt cutting inserts by 3DP. Such fabrication offers the potential of several degrees of flexibility over the current practice of dry pressing, including flexibility in geometry, in composition, and in response to the market demand.

6. *Metal Matrix Composites by 3DP*   The purpose of this research is to fabricate ceramic preforms for metal matrix composites. 3DP will be used to create the preforms and thereby to define the shape of the final part. Pressure infiltration used by metal matrix cast composites will be used to create the composites.

5.3   Z CORPORATION'S 3DP PROCESS (Z406 SYSTEM)

5.3.1   Company Profile

Z Corporation is a privately owned company with its headquarters in Massachusetts.* Z Corp. commercializes MIT's 3DP process of building 3D models from CAD files and other 3D CAD data. With the need for speed, and ease of use and quality in mind, a tool has been developed for engineers, marketers, and manufacturers to communicate and improve designs in three dimensions instead of working with two dimensions on a computer screen.

Although relatively new in the field of RP, Z Corp. has established itself as a viable player and developed a reputation for delivering the fastest RP machine available. Z Corp. is now the third-best reseller of RP machines in the world.

5.3.2   Process

System Hardware   Z Corp. incorporates the 3DP process into the Z406 system. The Z406 system is a premium 3D printer with the capability of printing in full color. This model is one of the fastest RP machines to make solid models. The Z406 can build models up to $8 \times 10 \times 8$ in. ($203 \times 254 \times 203$ mm). Parts built with original starch material can be hardened to fit the application necessary. Wax infiltration gives the parts some strength but also leaves them usable as investment casting patterns. Stronger infiltrants such as cyanoacrylate can be used to strengthen the part so that it can withstand rough handling. Some of the important components of the Z406 are discussed below [4]:

1. *Build and Feed Pistons*   These pistons provide the build area and supply material for growing parts. The build piston lowers the platform as part layers are printed, while the feed piston raises the platform to provide a layer-by-layer supply of new materials. This provides the motion in the z direction of the part build.
2. *Printer Gantry*   This printer gantry provides the x and y motion of the part building process. It also houses the print head, the printer cleaning station, and the wiper/roller for powder landscaping.
3. *Powder Overflow System*   This system is an opening opposite the feed piston where excess powder scraped across the build piston is collected. The powder is pulled down into a disposable vacuum bag both by gravity and an onboard vacuum system.

---

*Address: Z Corporation, 32 Second Avenue, Burlington, MA 01803. Telephone: (781) 852-5005; website: www.zcorp.com; e-mail: sales@zcorp.com.

4. *Binder/Take-up System*   The liquid binder is fed from the container to the printer head by a siphon technique, and the excess pulled through the printer cleaning station is drained into a separate container. Sensors near the containers warn when the binder is low or the take-up is too full.

Operation   The operation of Z406 is very simple, requiring very few commands. Since the parts are built in a powder bed, no support structures are required for overhanging surfaces.

Software   Like any other RP machine operation, the Z406 system starts with the standard binary STL file format, which is imported into the Z Corp. software. The file is then sliced into 2D cross sections and saved as a BLD (build) file. The software allows the part to be oriented at any angle and sized as required. Multiple STL files can be imported to build various parts simultaneously. The default slice thickness is 0.008 in. However, the value can be varied to meet the needs for particular parts.

After the STL file is imported, sliced, and oriented in the build envelope, 3D Print command is used to send the part file to the machine in order to grow the part. When the part is completed, a dialog box is displayed that shows the final build time, the total volume of material used, and the average droplet size of the binder used.

Machine Preparation for a Build   The Z406 machine must be checked and made ready before any part can be printed. The feed piston should have sufficient powder, and the build area needs to be scraped by the wiper blade until it is level with the powder. It in not necessary to check the container bag frequently since it lasts for several months. The vacuum bag, which collects the overflow powder, needs to be checked every few days. The jet cleaning station is squirted with distilled water to clean the print jets. All instructions regarding the safe and efficient use of the Z406 system should be known for the optimum performance of the system.

Principles   Before the printing of any part, the Z406 system software imports the STL file and slices it into hundreds of 2D cross sections of certain slice thickness. To print a part, the following steps are taken [4]:

1. The machine spreads a layer of powder from the source piston to cover the surface of the build piston.
2. The machine then prints binder solution onto the loose powder, forming the first cross section where the binder is applied, and the powder is glued together. The remaining powder remains loose and supports the layers that will be printed above.

3. When a cross section is complete, the build piston is lowered slightly, a new layer of powder is spread over its surface and the process is repeated.
4. The part grows layer by layer on the build piston until the part is complete while it remains surrounded and covered by loose powder.
5. Finally the build piston is raised and the loose powder is vacuumed away, exposing the completed part.

Table 5.4 provides a summarized specification of Z406 [4].

ZPrinter 310 System   The ZPrinter 310 System is the ideal entry-level rapid prototyping system that creates physical models directly from digital data in hours instead of days. The system is ideal for an office environment or educational institution, providing product developers easy access to a 3D Printer. Figure 5.8 shows a ZPrinter 310 system.

Spectrum Z510 System   The Spectrum Z510 Full Color System produces high-definition, full-color prototypes quickly and affordably. Superior injet printing technology creates parts with crisply defined features, enhanced accuracy, and precise color. This unique, 24-bit color, 3D printing capability produces color models that accurately reflect the original design data. Color models communicate more information than any other type of rapid prototype, providing the designer with a strategic advantage in product development. Figure 5.9 shows a Z510 system.

Z810 System   The Z810 System is the fastest and least expensive way to create large appearance prototypes for design review, mock-ups for form and fit testing, and patterns for casting applications. The large build volume allows

TABLE 5.4   Summarized Specification of Z406 (Z Corp. Inc.)

Process	3D Printing
Build volume (in. and mm)	8 × 10 × 8 in.
	(203 × 254 × 203 mm)
Layer thickness (in. and mm)	0.003–0.01 in.
	(0.076–0.254 mm)
Build speed (vertical)	Color mode: 2 layers per minute
	Monochrome mode: 6 layers per minute
Size of unit	29 × 36 × 42 in.
	(74 × 91 × 107 cm)
Power requirement	115 or 230 V, 50/60 Hz
Equipment weight	470 lb (210 kg)
Price (2001)	$33,500 U.S.

Figure 5.8 ZPrinter 310 system. (Courtesy of Z Corp., Inc.)

designers to print full-scale concept models for more effective communication with marketing, manufacturing, customers and suppliers. The Z810 System's color capability allows accurate representation of designs including FEA and other engineering data, further enhancing communication. Physical models can be created in plaster or starch-based materials and can be infiltrated to produce parts with a variety of material properties, satisfying a wide spectrum of modeling needs. Figure 5.10 shows the large Z810 system.

Applications of Parts
- Concept verification
- Form, fit, and function

Figure 5.9 Spectrum Z510 system. (Courtesy of Z Corp., Inc.)

Figure 5.10   Z810 system. (Courtesy of Z Corp., Inc.)

- Early detection of design error
- Investment or sand cast patterns
- Ability to be used in an office environment

Advantages and Disadvantages

Advantages

- Speed
- Material properties
- Color

Disadvantages

- Surface finish is not that smooth
- Frequent change of ink cartridges
- Low dimensional tolerance

5.4   SOLIGEN'S DIRECT SHELL PRODUCTION CASTING

5.4.1   Company Profile

Soligen, Inc.* was founded by Yehoram Uziel, its president and CEO, in 1991 and went public in 1993. It first installed its Direct Shell Production Casting

---

*Address: Soligen, Inc., 19408 Londelios Street, Northridge, CA 91324. Telephone: (818) 718-1221; website: www.soligen.com; e-mail: user@soligen.com.

TABLE 5.5

	ZPrinter 310 System	Spectrum Z510 System	Z810 System
Build size	8″ × 10″ × 8″	10″ × 14″ × 8″	20″ × 24″ × 16″
Layer thickness	User selectable at time of printing; .003″–.010″	User selectable at time of printing; 0.0035″–0.008″	User selectable at the time of printing: 0.003″–0.010″
Material options	High performance composite, snap-fit, elastomeric, direct casting, investment casting	High performance composite, direct casting	High performance composite, snap-fit, direct casting, investment casting
Equipment dimensions	29″ × 32″ × 43″	42″ × 31″ × 50″	95″ × 45″ × 76″
Equipment weight	250 lbs	450 lbs	1240 lbs
Equipment software	Z Corporation's proprietary software accepts solid models in STL, VRML and PLY file formats as input. ZPrint software features 3D viewing, text labeling, and scaling functionality. The software runs on Microsoft Windows* NT, 2000 Professional and XP Professional.	Z Corporation's proprietary software accepts solid models in STL, VRML and PLY file formats as input. ZPrint software features 3D viewing, text labeling, and scaling functionality.	Z Corporation's proprietary software accepts solid models in STL, VRML and PLY file formats as input. ZPrint software features 3D viewing, text labeling, and scaling functionality. The software runs on Microsoft Windows* NT, 2000 Professional and XP Professional.

(DSPC) system at three "alpha" sites in 1993. This technology generates three-dimensional molds that are then used for metal or plastic casting. *Soligen does not offer its rapid prototyping machines for sale. Rather it offers its services to create the mold and perform the casting from the CAD files submitted by the customers.* The ceramic molds created by the Soligen DSPC system allows for a wide range of metal alloys to be cast [5]. It can be said that DSPC is similar to rapid tooling, which will be discussed in Chapter 8.

5.4.2   Model and Specifications

The DSPC process creates ceramic molds for metal parts with integral coves directly and automatically from a CAD file. The specifications for DSPC 300 are shown in Table 5.6. Soligen's DSPC machine, the DSPC 300 (Fig. 5.11), includes the following mechanisms:

1. A holder that contains the manufacturing material—powder
2. A powder distributor to distribute a thin layer of powder
3. Rollers that are used to compress each layer before binding
4. A print head that sprays binder on each layer
5. A bin that is used to hold the mold

5.4.3   Key Strengths

The DSPC process is the only rapid prototyping process that creates ceramic molds for metal casting. As a result, functional metal parts (or metal tooling such as dies for die casting) are made directly from CAD data of the part. DSPC is a proprietary fabrication process for metal parts that combines the advantages of both casting and machining while eliminating the disadvantages of both processes. Figure 5.12 illustrates how DSPC uses only the advantages of conventional casting and machining in its product fabrication. The DSPC

TABLE 5.6   Summarized Specifications of DSPC 300 (Soligen Inc.)

Model	DSPC 300
Process	Direct shell production casting
XY resolution (mm)	0.05
Work volume, XYZ (mm × mm × mm)	304 × 304 × 304
Layer thickness (mm)	0.178
Vertical build rate (mm/hr)	12.7–19.0
Size of unit, XYZ (m × m × m), estimated	2.3 × 1.3 × 2.6
Data control unit	EWS
Power supply	110 or 220 V_{AC}, 10 A
Price (US$, 1996)	250,000

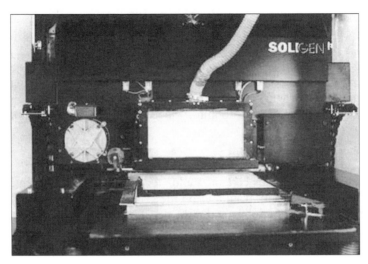

Figure 5.11   Soligen DSPC 300 rapid prototyping machine. (Courtesy of Soligen, Inc.)

process significantly reduces product development lead time by eliminating many steps in the conventional casting procedure. Figure 5.13 compares the conventional casting process with that of the DSPC process. This convention casting flow diagram does not include a block for rework. Rarely does an original model go through the development process without receiving modifications. Using the conventional method, a modification takes place at the part design stage, which is the first step of the process. Modifications to the

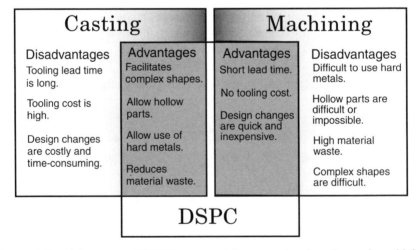

Figure 5.12   Advantages of DSPC compared to conventional casting and machining processes. (Courtesy of Soligen, Inc.)

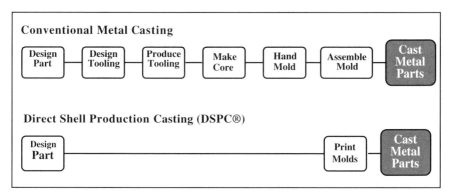

Figure 5.13 Conventional versus DSPC casting processes.

CAD model used in DSPC can be made in real time and do not require any costly changes to tooling or added machine time. Using DSPC allows for the tooling to be designed and fabricated after the final design has been tested and ready for mass production. Table 5.7 [5] compares the cost and lead time of conventional casting and the DSPC process. The individual part fabrication cost using DSPC is much greater then using the conventional method, but, when the tooling cost is factored in, DSPC is roughly seven times less expensive.

5.4.4 Process

The principle of Soligen's DSPC process is based on 3DP a technology invented, developed, and patented by MIT. 3DP is licensed exclusively to Soligen on a worldwide basis for the field of metal casting. In the process, parameters that influence performance and function are the layer thickness, powder's properties, the binders, and the pressure of rollers. The process (see Fig. 5.14) contains the following steps [5]:

1. A part is first designed on a computer, using a commercial CAD software.

TABLE 5.7 Comparison of Options

	Conventional Permanent Iron Mold	Hybrid	DSPC RP System
Tool cost	$80,000–$100,000	$25,000–$35,000	$0
Piece price	$100–$150	$1,000–$1,500	$10,000–$14,000
Lead time	10–12 weeks	4–6 weeks	15 days

1. Customer file is received.
Design a part of virtually any shape and send the file or drawing to Soligen. Here the part is a hollow turbine blade used in a jet engine.

2. Gating is added.
Soligen first designs a virtual pattern for net shape casting, including a gating system through which molten metal will flow.

3. Ceramic mold is converted.
A digital model of the mold and is created and transfered the model to the **DSPC** system. The mold includes integral cores to produce hollow sections.

4. Ceramic mold is built.
The **DSPC** system automatically generates the mold from thin layers of ceramic powder. After lowering the previous layer, a new layer of powder is spread.

a.
The liquid binder is"printed" onto the powder layer to define a cross section of the mold.

b.
The process is repeated until the entire mold is printed. This is then fired, resulting in a rigid ceramic mold surrounded by unbound powder.

c.
The unbound powder is removed from the mold

1.0   The mold is Cast
The last step is to fill the mold with molten metal at our foundry. After the metal solidifies, the ceramic and gating metal are removed and the part is finished, inspected and shipped.

Figure 5.14   DSPC process. (Courtesy of Soligen, Inc.)

2. The CAD model is then loaded into the shell design unit, the central control unit of the equipment. Preparing the computer model for casting mold requires modifications such as scaling the dimensions to compensate for shrinkage, adding fillets, and removing characteristics that will be machined later. The mold maker then decides how many mold cavities will be on each shell and the type of gating system, including the basic sprues, runners, and gates. Once the CAD mold shells are modified to the desired configuration, the shell design unit generates an electronic model of the assembly in slices to the specified thickness. The electronic model is then transferred to the shell production unit.

3. The shell production unit begins depositing a thin layer of fine alumina powder over the shell-working surface for the first slice of the casting mold. A roller follows the powder, leveling the surface.

4. An ink-jet print head, similar to those in computer printers, moves over the layer, injecting tiny drops of colloidal silica binder onto the powder surface from its 128 ink jets (Fig. 5.15). Passing the pressurized stream of binder through a vibrating piezoelectric ceramic atomizes it as it exits the jet. The droplets pick up an electric charge as they pass through an electric field that helps to align them to the powder. The binder solidifies the powder into ceramic on contact and the unbonded alumina remains as support for the following layer. The work platform lowers and another layer of powder is distributed.

5. The process through steps 3 and 4 is repeated until all layers of the mold have been formed.

6. After the building process is completed, the casting shell remains buried in a block of loose alumina powder. The unbound excess powder is then separated from the finished shell. The shell can then be removed for postprocessing, which may include firing in a kiln to remove moisture or preheating to an appropriate temperature for casting.

7. Molten metal can then be poured in to fill the casting shell or mold. After cooling, the shell can be broken up to remove the cast that can then be processed to remove gatings, sprues, and the like, thus completing the casting process.

The hardware of the DSPC system comprises a PC computer, a powder holder, a powder distributor, rolls, a print head, and a bin. The software includes a CAD system and a Soligen's slicing software.

5.4.5 Applications and Uses

The DSPC technology is used primarily to create casting shells for production of parts and prototypes. Soligen Inc. has acquired a foundry in Santa Ana, California, for commercial production of prototypes and molds. Called Parts

Figure 5.15 Multijet print head.

Now Division, it serves as a service center as well as a place for equipment and process research and development (Fig. 5.16). It aims to be a premier "one-stop shop" for functional cast metal parts produced directly from a CAD file and with no need for prefabricated tooling to produce the first article. DSPC has been used in the following areas:

1. Automotive industry
2. Aerospace industry
3. Computer manufacture
4. Medical prostheses

The Parts Now program includes [5]:

- Fast turnaround of fully functional metal part(s) delivered in days before any tooling (pattern or core box) could be made or even designed. These parts are cast directly from the customer CAD file and enable design iterations to be fully implemented and tested, bypassing the lead time and cost of cast tooling.
- DSPC-made hybrid-casting tools: Concurrent with the testing progress and as parts of the design are "frozen for production," hybrid tools are made. These include conventional patterns with "drop-in" DSPC-made sections and/or cores. The hybrid tools approach enables larger quantities of parts for testing. It also allows modifying sections of the parts under development without disposing of the casting tools. Hybrid tools progress with the part's development and become conventional production tools as the development and testing are successfully completed.
- Soligen's Parts Now service combines DSPC with traditional casting and CNC machining to provide a seamless transition from a CAD file to mass production manufacturing.

Figure 5.16 Because of DSPC, and a successful Parts Now program, the timely production of this manifold resulted in a win at the Daytona 500 in early 1999.

- Launching mass manufacturing: Concurrent with the development process a casting process and appropriate mass manufacturing foundry are identified. The Soligen team in collaboration with the production foundry produces mass production casting tools. Soligen provides all the engineering and first article certification of the mass production line.
- A Parts Now program ensures that the first prototype and the 10,000th production part are identical.
- A smooth transition from CAD to conventional casting.

5.4.6   Case Studies

Soligen's Parts Now Pilots Metal Parts for Caterpillar   Caterpillar needed to prototype a complex engine component within a week [5]. It sent the design file of the component via a modem to Parts Now (a division of Soligen Inc.) on Wednesday evening. By Friday, Parts Now completed two casting shells with the DSPC systems. On Monday, the foundry poured the shells with A356 aluminum. On Tuesday, the parts were heat treated and finished machining on Wednesday. On the same day, the fully functional parts were shipped to Caterpillar for installation on the engine, thus completing the order from Caterpillar. This is very much faster than any traditional fabrication process could deliver.

5.4.7   Research and Development

Soligen Inc. continues to conduct research on part and mold design. Using knowledge and experience from application development, the Parts Now Division is developing technology to produce intricate aluminum die casts that could improve throughput by 50 to 250% while improving accuracy and durability as well. Improving mold surface finishes with DSPC-cast electrode used with electric discharge machining (EDM) and coating technologies are also under development.

5.5   EOS'S LASER SINTERING SYSTEMS

5.5.1   Company Profile

Electro Optical System (EOS)* was founded in 1989 and was the first European manufacturer of laser-based rapid prototyping systems. EOS launched the STEREOS laser-stereolithography system in 1991 with a liquid-based laser modeling system. Three years later it produced the EOSINT P, which was the beginning of its EOSINT product line. In 1997, EOS exited the

---

*Address: Electro Optical Systems, Robert-Stirling-Ring 1, D–82152 Krailling/Munich, Germany. Telephone: +49 (0)89 893 36-0; website: www.eos.info; e-mail: info@eos.info.

stereolithography technique and focused entirely in Laser Sintering technology. The EOSINT product line is a powder-based laser sintering system that has three different technologies: EOSINT P, EOSINT M, and EOSINT S [6].

EOS is the leading European supplier for laser sintering systems and has made developments in process chains, systems, materials, and software for industrial applications.

5.5.2 Models and Specifications

EOSINT P EOSINT P systems build plastic parts from polyamide or polystyrene directly from CAD data, without support structures. The EOSINT P systems come in two models: the P 380 (see Fig. 5.17) and the P 700 (see Fig. 5.18), depending on the production size. The systems allow for the production of fully functional parts ranging from a size of $340 \times 340 \times 620$ mm (see Table 5.8) to $700 \times 380 \times 580$ mm (see Table 5.9). Through exposure strategies and process control, they offer a very high building speed and excellent part quality. Further parts to be built can be loaded during the building process as well. High process integration and automation guarantee minimum turnaround times. EOSINT P systems thus combine flexible rapid technology with the automation and efficiency of mass production [6].

EOSINT M The EOSINT M 250 system builds metal tooling for injection molding, die casting, sheet metal forming, and vulcanizing, as well as metal parts, directly from steel-based and other metal powders. The EOSINT M 250

Figure 5.17 EOSINT P 380 laser sintering system. (Courtesy of EOS.)

Figure 5.18 EOSINT P 700 laser sintering system. (Courtesy of EOS.)

TABLE 5.8 Specifications for EOSINT P 380

Effective building volume	$340 \times 340 \times 620$ mm ($B \times D \times H$)
Building speed (material dependent)	10–25 mm height/hr
Layer thickness (material dependent)	0.15 mm
Laser type	50 W, CO_2
Precision optics	F-theta-lens
Scan speed (maximum)	5.0 m/sec
Power supply	32 A
Power consumption (maximum)	4 kW
Nitrogen generator	Integrated (optional)
Compressed air supply	Minimum 5000 hPa; 6 m³/hr
Dimensions	
Process cabinet	$1250 \times 1300 \times 2150$ mm ($B \times D \times H$)
Control terminal	$610 \times 820 \times 1785$ mm ($B \times D \times H$)
Recommended installation space	$4700 \times 3700 \times 3000$ mm ($B \times D \times H$)
Weight	Approx. 900 kg
Data preparation	
PC	Current Windows operating system
Software	EOS RP tools; Magics RP (Materialise); Expert Series (DeskArtes)
CAD interface	STL, CLI
Network	Ethernet
Certification	CE

TABLE 5.9 Specifications for EOSINT P 700

Effective building volume	700 × 380 × 590 mm ($B \times D \times H$)
Building speed (material dependent)	10–25 mm height/h
Layer thickness (material dependent)	0.15 mm
Support structure	Not necessary
Laser type	50 W, CO_2
Precision optics	F-theta-lens
Scan speed (maximum)	5.0 m/sec
Power supply	32 A
Power consumption (maximum)	2.2 kW
Nitrogen generator	Integrated
Compressed air supply	Minimum 6000 hPa; 20 m^3/hr
Dimensions	
System incl. switchgear cabinet	2270 × 1410 × 2100 mm ($B \times D \times H$)
Control terminal	850 × 750 × 1630 mm ($B \times D \times H$)
Powder supply system	1700 × 1050 × 1320 ($B \times D \times H$)
Break-out station	1600 × 800 × 1080 mm ($B \times D \times H$)
Recommended installation space	6200 × 4600 × 3000 mm ($B \times D \times H$)
Weight	Approx. 2300 kg
Data preparation	
PC	Current Windows operating system
Software	EOS RP Tools; Magics RP (Materialise); Expert Series (DeskArtes)
CAD interface	STL; optional: converter to all common formats
Network	Ethernet
Certification	CE

(see Fig. 5.19) uses technology known as direct metal laser sintering (DMLS) in order to produce parts such as tool inserts as well as prototypes and final products directly from 3D CAD data. The EOSINT M 250 has a building volume of 250 × 250 × 200 mm (including building platform) (see Table 5.10) [6].

EOSINT S The EOSINT S 750 (see Fig. 5.20) system builds cores and molds directly from CAD data for the production of sand castings without any tooling. EOSINT S uses a double laser sintering system that allows for processing of Crouning molding material. High-quality castings can be produced in fields such as motor development, pumps, or hydraulic applications. This system produces lightweight constructions using aluminum or magnesium and provides new applications for cast iron and steel. The technical specifications for EOSINT S 750 are shown in Table 5.11.

Figure 5.19 EOSINT M 250 laser sintering system. (Courtesy of EOS.)

TABLE 5.10 Specifications for EOSINT M 250

Maximum building volume	$250 \times 250 \times 200$ mm (including building platform)
Building speed (material dependent)	$2-15$ mm^2/sec
Layer thickness (material dependent)	$20-100$ μm
Laser type	Min. 200 W, CO_2
Precision optics	F-theta-lens, high-speed scanner
Scan speed (maximum)	3.0 m/sec
Power supply	32 A
Power consumption (maximum)	6 kW
Nitrogen generator	Integrated (optional)
Compressed air supply	Minimum 7000 hPa; 6 m^3/hr
Dimensions	
System	$1950 \times 1100 \times 1850$ mm ($B \times D \times H$)
Recommended installation space	$4.7 \times 3.7 \times 3.0$ m ($B \times D \times H$)
Weight	Approx. 900 kg
Data preparation	
PC	Current Windows operating system
Software	EOS RP tools; Magics RP (Materialise); Expert Series (DeskArtes)
CAD interface	STL optional converter for all standard formats
Certification	CE

Figure 5.20   EOSINT S 750 laser sintering system. (Courtesy of EOS.)

TABLE 5.11   Specifications for EOSINT S 750

Effective building volume	$720 \times 380 \times 380$ mm
Building speed (material dependent)	Up to 2500 cm³/hr
Layer thickness	0.2 mm
Laser type	2×100 W, CO_2
Precision optics	$2 \times$ F-theta-lens, $2 \times$ high-speed scanner
Scan speed (maximum)	3.0 m/sec
Power supply	32 A
Power consumption	6 kW
Compressed air supply	Minimum 6000 hPa; 15 m³/hr
Dimensions	
Process cabinet	$1400 \times 1400 \times 2150$ mm ($B \times D \times H$)
Control cabinet	$610 \times 800 \times 1830$ mm ($B \times D \times H$)
Switchgear cabinet	$810 \times 870 \times 2150$ mm ($B \times D \times H$)
Recommended installation space (without IPCM)	$4.5 \times 4.6 \times 2.7$ m ($B \times D \times H$)
Weight (without sand)	Approx. 2,200 kg
Data preparation	
PC	Current Windows operating system
Software	EOS RP tools; Magics RP (Materialise); Expert Series (DeskArtes)
CAD interface	STL optional converter for all standard formats
Network	Ethernet
Certification	CE

5.5.3   Laser Sintering Process

All EOSINT systems are based on the principle of laser sintering. The process uses a laser beam to create a solid part directly from 3D CAD design, which is sliced into thin layers in the software. The system disperses powder and is followed by a coating in order for the sintering reaction to occur as each layer is created. A laser beam is used to create a physical replica of the 3D CAD model layer by layer from fine powder. The platform lowers with each created layer. The layer thickness ranges from a few tenths to one hundredths of a millimeter depending on the application. The cycle is repeated until the entire part is created.

5.5.4   Applications and Uses

Rapid product development with EOS technologies further enhances simultaneous engineering, as prototypes and tools can be created throughout the entire product development process. As a result, individual phases can be accelerated even further. At the same time, development quality improves since errors can be avoided or detected in earlier stages before incurring major costs or endangered product launch [6]. Table 5.12 shows the applications of EOS technologies for both plastic and metal parts.

TABLE 5.12   Applications of EOS Technologies for Plastic and Metal Parts

	Plastic Parts
DirectPart	Used for styling and design or creating functional prototypes such as steering wheel, control panels, fuel tank, ventilation system, and car console
DirectPattern	Used in making functional prototypes via silicone rubber molding
DirectTool	Used for creating technical prototypes: a component manufactured from the target material using series production methods and standard parameters.

	Metal Parts
DirectPart	Used for styling and design or creating functional prototypes such as toothed belt wheel or motor housing.
DirectPattern	Used for creating technical prototypes and small series of investment cast parts such as an axle. Also used for making functional prototypes for die-cast and sand-cast parts.
DirectTool	Used for creating technical prototypes and small series of die-cast parts such as window connectors and clutch housings.
DirectCast	Used in making technical prototypes and small series of sand-cast parts such as cylinder heads, hydraulic controllers, impellers, and steering blocks.

5.6   e-MANUFACTURING USING LASER-SINTERING

e-manufacturing means the direct, flexible and cost effective production directly from 3D CAD data files. Laser-sintering is a very efficient way of making e-manufacturing reality, because the 3D CAD can directly be fed into a laser-sintering machine. The result of an e-manufacturing process can either be the final product or a tool that facilitates manufacturing the final product. When considering laser-sintering as a means to implement e-manufacturing, mainly final products of plastics or metal can be achieved. While plastic parts can directly result from a laser-sintering process, metal parts can either be directly sintered or be facilitated by sintering a mold, a tool or a cast. While molds and tools are another application for direct metal laser sintering, casts can be produced by laser-sintering of sand. Sacrificed models can be made e.g. of polystyrene processed on a laser-sintering system for plastics. These polystyrene or sand die casts eventually result in metal parts.

5.6.1   Introduction

A substantial shift from mass marketing to customer segment marketing has taken place, and is presently evolving into the form of personalized marketing [7]. The ability to transfer an in-depth knowledge of what a customer wants, and how much they will pay for it, into a physical product and service has allowed this paradigm shift. This results in the ability to offer individualized products and services in the sense of mass customization. This requires both restructuring efforts in sales and manufacturing [8]. While the efforts in sales focus around minimizing the customer perceived uncertainty when acquiring customized products, research in manufacturing revolves around the issue of producing customized products at costs comparable to series production. In this chapter we will investigate what contribution e-manufacturing can make in this sense. From our point of view, e-manufacturing means the direct, flexible and cost effective production directly from 3D CAD Data files. In this way laser-sintering is a very efficient way of making e-manufacturing a reality.

As a rule of thumb we can claim that e-manufacturing has a good chance of unlaishing its potential if

1. The customer demands a high number of varieties which culminates in the extreme in the demand for individual products.
2. Demanded products are of high complexity.
3. Customer demand is subject to sudden and unpredictable change.
4. Product life-cycles ever shorten.
5. The sheer number of identical products sold is low.

The result of e-manufacturing activities can be the following:

- Direct parts as ready-to-use products or components within customized products
- Direct tools that support conventional manufacturing processes
- Direct casts or patterns as sacrificed cores and molds

In this section, we will give a number of examples that prove that e-manufacturing with laser-sintering can be a viable solution for industries that fit the above defined rule of thumb. All these case-studies have been implemented on EOSINT equipment. The initiators of these case-studies were either EOS, independent service bureaus, the end customers or a consortium consisting of these parties.

5.6.2 Customized Plastic Parts

Sunglasses Sunglasses are a prime example for a product that is offered in sheer unlimited variations. While the product complexity can certainly be handled with, the demand for sunglasses is very much subject to current trends and fashion styles. This makes it very hard for sunglass manufacturers to predict a plausible forecast for next years sales. If the manufacturer over-estimates demand this might result in excessive stocking costs and heavy discounts to make the overstocked products sell. However, the other extreme, the underestimation of future demand, isn't any better. The inability to offer customers what they want results in lost sales. This dilemma calls for a production method that easily can adapt to both a change in fashion trends and in demanded quantity. Laser-sintering can be such a method as shown in this case study.

Implementing e-manufacturing also can open new ways in sales. In this case it very much could be possible that the customer enters a kind of "mini-factory" where she could design her own sunglasses at a suitable "tool-kit" providing her with support in doing so. Once her design efforts result in sunglasses she will be happy to buy, the created 3D data is directly transferred to a EOSINT system where the sunglasses are built. In order to transfer the crude sintered part to a posh fashion product it still has to undergo a AutoFinish process that results in sunglasses as shown in Fig. 5.21.

Hearing Aids Another industry that holds true for our defined rules of thumb are hearing aids. Since the success of hearing aids very much depends on its ability to adapt to the anatomy of the auditory canal. Fig. 5.22 illustrates a hearing-aid with a laser-sintered shell.

The process of e-manufacturing a hearing aid is as following:

Figure 5.21   Laser-sintered sunglasses. (Courtesy of Tecnologia & Design, Crabbi Sunliving, EOS.)

1. Take a copy of the auditory canal anatomy by creating a wax cast.
2. Scan the wax cast with a scanner (e.g., Steinbichler) to create 3D data.
3. Integrate an identification number in the 3D data that helps identify the product after the laser-sintering process.
4. Laser-sinter the shell.
5. Combine the laser-sintered shell with the electronics components.

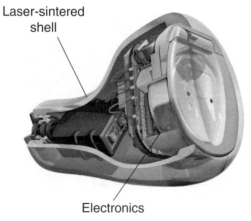

Laser-sintered
shell

Electronics

Figure 5.22   Laser-sintered hearing aid. (Courtesy of Phonak.)

Skiboot Buckles   Figure 5.23 illustrates the result of a laser-sintered and auto finished skiboot buckle that was painted in the last process steps.

Whether skiboot buckles, inline skaters, ice-hockey boots or any other trendy sports shoe, the buckle can become a vital part in determining both individual fit and aesthetic preferences. The major benefit of e-manufacturing buckles like these with laser-sintering equipment is in a significantly reduced time-to-market. Because of omitting tooling that is necessary in traditional manufacturing (e.g., blowmolding) the entire manufacturing process is sped up. Figure 5.24 illustrates the time saving effects of up to 58%.

A T&D research program in 2004 on this particular product indicated that laser-sintering leads blowmolding from an economic point of view up to 630 units [9]. For many consumer goods industries this number is indeed very hard to exceed, making e-manufacturing with laser-sintering the manufacturing method of choice.

Public Phone Cradle   Leaving consumer goods, e-manufacturing can be even more valuable for certain B2B applications. A very good example is the e-manufacturing case from Model Shop Vienna, using laser-sintering technology to e-manufacture a cradle for a public phone receiver (Fig. 5.25).

The issue for Model Shop Vienna was to produce a lot size of 250 of the described cradles (Fig. 5.26). Since this lot size is quite limited, conventional manufacturing methods were not suitable. Heavy investments in tooling would have increased cost per unit dramatically. In addition, once tooling was fixed, design changes would be prohibitive. Since Model Shop could not say for sure how many units of this cradle they would have to build, they decided to go for laser-sintering with Alumide (which is an EOS trademark). This material combines all the properties the customer Telekom Austria demanded (metallic appearance, high stiffness, etc.). Also, laser-sintering gave Model Shop Vienna the flexibility to respond to unpredictable design changes the

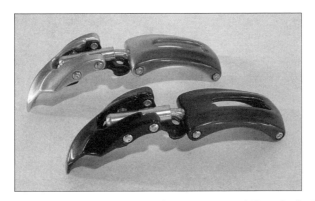

Figure 5.23   Laser-sintered skiboot buckles. (Courtesy of Tecnologia & Design.)

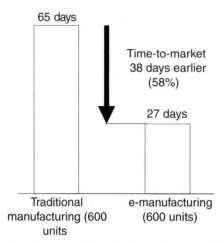

Figure 5.24 Comparing time-to-market of a laser-sintered skiboot buckle with traditional manufacturing methods.

customer might want to add in a second order. Since laser-sintering results in a linear cost curve, Model Shop Vienna did not have to calculate the break-even-point where the heavy investments in tooling would pay off. Model Shop Vienna estimates that laser-sintering provides a cost advantage compared to conventional manufacturing at quantities of up to 500 identical pieces.

Hammtronic Joystick "HI-Drive" Another impressive example for e-manufacturing is the manufacturing of a joystick used to navigate a Hammtronic tandem roller. By means of an automatic translation, the vehicle always moves

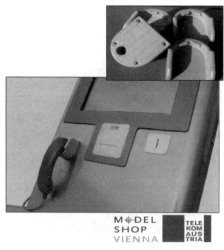

Figure 5.25 Laser-sintered public phone cradle.

Figure 5.26 Fitting 120 cradles into build window. (Courtesy of Model Shop Vienna.)

in the direction of the joystick deflection depending on which way the driver is sitting. All in all this joystick offers 16 different functions and consists of 15 plastic parts plus parts for electronics, mechanics and switches.

With a lot size of approximately 5,000, the advantages of e-manufacturing were not in laser-sintering the components themselves but in laser-sintering the metal tools for injection molding the components. The first step was to divide the parts into four master tools for optimum utilization of the injection molding machine. Then minimization of the volume of the laser-sintered inserts was required for minimum build times and material consumption. This resulted in "onserts" instead of inserts that were attached to a conventionally manufactured tool base. The total volume of laser-sintered material was 3.1 liters for all tooling for the 14 plastic parts. Figure 5.27 illustrates some of these inserts.

The major value added process in e-manufacturing these inserts with laser-sintering was in a dramatically reduced time to market. In this particular case, the Hammtronic was about to be presented at a trade show just a few months ahead. Since the injection molding process was not yet optimized the tools underwent certain iterations until the molding process was optimized. Laser sintering provided the flexibility to create these iterations at significantly reduced costs.

Rubber Boots   A very interesting example showing the potential of direct tooling in plastics laser-sintering is the injection molding of a rubber boot sole as illustrated in Fig. 5.28 [10].

Figure 5.27 Some of the insert geometries laser-sintered on EOSINT M. (Courtesy of Fruth Innovative Technologien—FIT.)

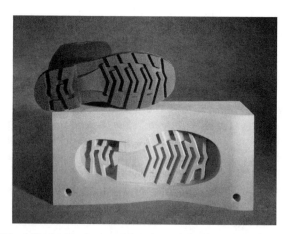

Figure 5.28 DirectTool injection mold and resulting rubber boot sole. (Courtesy of Tecnologia & Design.)

The advantage of employing laser-sintering in this case is primarily the reduced time-to-market. Following the previous example of e-manufacturing the boot buckles, this method illustrates the direct way from 3D CAD data to an injection mold tool featuring the sole of the rubber boot or other trendy shoe. Using EOS Alumide further has the advantage that it can be used as a functional tool in the injection molding machine as is. This is possible because the material properties resemble that of metal. This also results in a tool endurance that is sufficient for producing several hundred shoe soles. This number is sufficient to produce trendy products like fashion shoes where design aspects are integrated into the sole design. Laser-sintering the tooling for this kind of shoe can help this industry adapt sole design to sudden changes in customer preferences quickly and at relatively low cost.

Multi-Component Molding   The superior quality of laser-sintered tools can also be demonstrated by the following example of a two color key ring used for a motor scooter. Figure 5.29 illustrates the process of getting a colored key ring using the DirectTool method employing direct metal laser-sintering.

In this tool, the parting surfaces fit perfectly. Only the reaming of ejector holes and polishing of molding surfaces was necessary after the laser-sintering process. The entire process from 3D CAD data to a finished tool took less than 2 days. In the manufacturing process the tools withstood 10,000+ high quality molded parts. This case shows that e-manufacturing with laser-

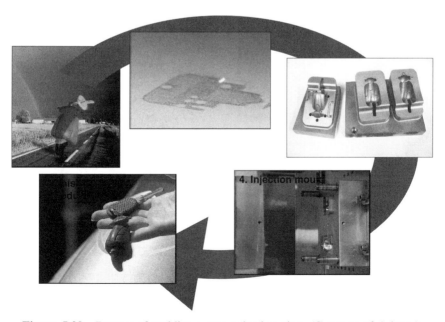

Figure 5.29   Process of molding a two color key ring. (Courtesy of Arberg.)

sintering is very well suitable to an application for even relatively large production batches. The quality of the tools can be demonstrated by the fact that a twin color mold requires very well finished tools as otherwise the two colors would bleed together.

Laser-Sintered Tool Inserts in Bottle Blow Molding   It is not always necessary or advantageous to manufacture entire tools with direct metal laser-sintering. Small inserts can be laser-sintered and the rest of the tool manufactured in a conventional way. This was the case in the following study where the blow molding of bottles was significantly improved by employing laser-sintered tooling inserts. Figure 5.30 illustrates where the laser-sintered tool insert was positioned within the bottle blowmolding tool.

The reason for utilizing the e-manufacturing process for this application was productivity improvement in series production of blow-molded PE bottles. The solution was a combination of conventional production tooling with inserts built by direct metal laser-sintering. The key innovation in this case was to integrate conformal cooling which is easy to do with laser-sintering but impossible with conventional manufacturing methods. The result was that cycle time was reduced from typically 15 seconds to just 8–9 seconds. This translated to a productivity increase to ~75%.

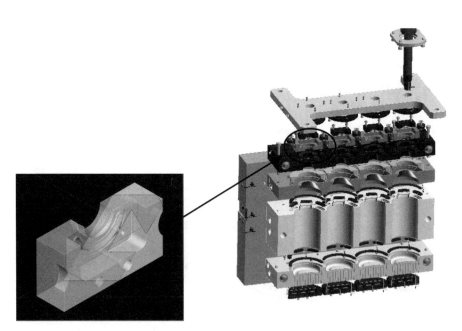

Figure 5.30   Illustration of laser-sintered tool insert position. (Courtesy of SIG Blow-tec, Es-Tec, DemoCenter.)

5.6.3 Customized Metal Parts

Dental Camera A good example for a metal part directly laser-sintered from 3D CAD data is the Planmeca dental camera as illustrated in the following Fig. 5.31.

The market for dental cameras is on the one hand relatively small and on the other hand lends itself to customization to the anatomy of the dentists' hands which adds significant utility and value to the user. This makes the customized manufacturing of dental cameras as one-offs feasible from an economic point of view. By doing so, manufacturers of such equipment can both address new markets by offering customization and gain an image boost by leveraging innovative technology.

Metal Spare Parts Another prominent example where e-manufacturing of metal parts with laser-sintering offers promising potential is the manufacturing of spare parts. Gallagher, Mitchke, and Rogers quantify the supply of after-market parts as a $400 billion business [11]. With the traditional manufacturing of spare parts there are several problems. These are described by the following:

- Spare parts must be provided for a long time even after the average product life-cycle has ended.
- Tooling, engineering documentation and the like must be retained for a substantial amount of time.
- Restarting the original manufacturing process (e.g., blow molding) is prone to problems for on-off production of spare parts.
- There are spare parts applications where no original tooling nor documentation exists (e.g., ancient automobiles, museum artifacts, etc.)

One viable solution to all of these problems is e-manufacturing of spare parts with laser-sintering. A real world example exists in many developing

Figure 5.31 Dental camera. (Courtesy of Planmeca.)

countries today as in a fleet of trucks that serve for many years after they have been replaced as obsolete in industrialized nations. If you were the African operator of a MAN 15.215 DHS truck, build in 1968, you certainly would experience great difficulties replacing an aging pump housing with original spare parts such as illustrated in the following Fig. 5.32.

Employing reverse engineering tools that generate 3D data from the pump housing to be replaced may be the only cost effective way to obtain a much-needed part. This data serves as the basis for e-manufacturing on laser-sintering systems which may be the only way to produce a spare part that meets the requirements.

The idea of e-manufacturing spare parts also becomes a very interesting solution for military purposes. In this field researchers evaluate the feasibility of a mobile parts hospital.* Here the scenario is that required spare parts are directly manufactured in the field. This scenario may also be tempting for commercial purposes. Imagine for example an ocean cruiser that e-manufactures required spare parts while at sea. The same might hold true for aerospace applications as spare parts could be e-manufactured on site at remote airports or even on a space station.

Computer Locks The requirement in this case study was the production of 300 sets of computer locking parts, each comprising assembly of 3 components as illustrated in Fig. 5.33.

The solution in this case was a DirectPart solution on an EOSINT M system using DirectMetal 20 powder. The required 300 sets were built in 5 jobs, 60 sets per job. The build time per job was 32 hours. This means that the required 300 sets were built in approximately 7 days. Although the time requirement of 7 days seems long at first sight it should be considered that the first 60 sets were built immediately after the 3D CAD design phase was

Figure 5.32 Imaginary example for a spare pump housing.

---

*Anonymous (2003): Mobile Parts Hospital. U.S. Army Tank-automotive and Armamants Command (TACOM), http://www.mobilepartshospital.com/welcome/, access date: 10.03.05.

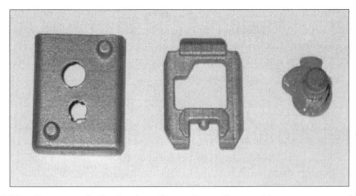

Figure 5.33   Computer lock: the three laser-sintered assembly components. (Courtesy of Rapid Product Innovations.)

over. With any conventional method of manufacturing, tools would have been necessary. The design and fabrication of those tools would easily have consumed much more than 7 days time.

Helicopter Turbine Parts   This case study was aimed at investigating e-manufacturing of rotors and stators for test rigs running at 30,000 rpm and 250°C (500°F). The stator as illustrated in Fig. 5.34 is for use in a MAKILA 2A helicopter engine. Its dimensions are 180 mm (7 inch) in diameter and 52 mm (2 inch) height. The blade thickness is between 0.5–1.8 mm (0.02–0.07 inch).

Figure 5.34   Helicopter Stator ring. (Courtesy of Pôle Européen de Plasturgie.)

The solution in this case was direct metal laser-sintering with DirectSteel 20. This resulted in a reduction of both time and costs by 50% compared to a conventional manufacturing method.

Although this is not an example for an end product it still demonstrates the magnitude of both e-manufacturing in the product design phase and the method of direct metal laser-sintering.

Hydraulic Valve for Rail Vehicle   When designing a new hydraulic valve for a rail vehicle the company approached ACTech, an EOS customer. The initial goal was to manufacture two prototypes to see whether the design was sufficient. The product is illustrated in the following Fig. 5.35.

To produce these prototypes ACTech laser-sintered the cores that were used in a follow on casting process. The unconventional way of producing cores however, had the positive side effect that the mold draft was not necessary anymore. This resulted in the switching properties of the prototype valve exceeding expectations. Therefore, the rail car company insisted on producing the entire batch size of 200 with laser-sintered sand cores. This order motivated ACTech to further pursue laser-sintering technology and succeeded in simplifying tooling complexity by using the single piece sand core and thus dramatically simplifying the outer casting molds. The result not only was a hydraulic valve with unprecedented switching properties but also an e-manufacturing process that reduced costs.

Racing Gearbox   Many interesting examples for e-manufacturing stem from the racing industry, led primarily by formula one racing teams. This not only is because of higher racing budgets in this class but also because of an inherent drive for any innovation that can shave another tenth of a second from lap times that may result in winning the race. The example presented here for a DirectPattern application was the requirement to produce a newly designed motor cycle gear box in titanium. Fig. 5.36 shows both the sacrificed

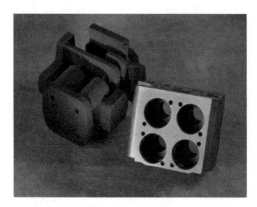

Figure 5.35   Laser-sintered sand core and end-product. (Courtesy of ACTech.)

Figure 5.36 Laser-sintered PrimeCast sacrificed model and final gear box. (Courtesy of Poggipolini.)

pattern laser-sintered with EOS PrimeCast 100 and the final titanium cast product.

In this application the DirectPattern was used as a sacrificial model in investment casting. Therefore the DirectPattern was coated with a ceramics based material that later formed the cast. After finishing, the DirectPattern was heated and melted out leaving the empty cast ready to be filled with the titanium alloy. The entire process was so fast that the originator Poggipolini Titanium coined the term "Racing parts at racing speed".

5.6.4 e-Manufacturing in a Brief

This section has demonstrated with numerous examples that e-manufacturing with laser-sintering can be a viable solution in a number of industries and manufacturing applications. It is important to stress that this not only holds true in a rapid prototyping environment but also in a series production environment of complex products that are offered in a high number of varieties. e-manufacturing gains the most impact in those cases where end-products can be manufactured with laser-sintering technologies either in plastics or in metal. By eliminating time and cost consuming detours through tooling processes, laser-sintering end-products can help certain companies gain a competitive edge over competitors including overseas manufacturers that have only a fraction of resident labor costs.

Also the other major application of laser-sintering as a DirectTool, DirectPattern or DirectCast application is a very cost effective solution in such cases where conventional methods such as injection molding are the preferred method of manufacture. As the case studies above showed, laser-sintering can help boost economic aspects of a product by providing fast access to high quailty tools. This concept gains even more momentum if the required tools are of great complexity thus leveraging the effect of laser-sintering and e-manufacturing. As the example of the Hammtronic joystick illustrated, producing tools by laser-sintering helps save precious time in the development process.

To sum up we can conclude that laser-sintering not only is a viable tool for rapid prototyping but increasingly becomes the manufacturing method of choice in industries that face a high number of varieties of complex products or added value by offering high levels of customization.

5.7   SUMMARY

In this chapter, the principles, operations, and applications of powder-based rapid prototyping systems were introduced and discussed. We have described the most established powder-based systems such as 3D Systems' selective laser sintering, MIT's three-dimensional printing, and EOS's laser sintering system. In addition, we have also described Z-Corp.'s 3D printing, Soligen's direct shell production casting, which utilizes MIT's 3D printing process, and EOS' laser sintering process.

3D Systems offers different SLS models to produce accurate and functional parts for many aspects of product development. The SLS system creates parts and physical prototypes quickly by skipping the tooling step. It builds durable metal parts and mold inserts directly from CAD files without casting and machining.

3D System offers Vangaurd and Vangaurd HS (high speed), the fourth-generation rapid prototyping machines. These machines are specifically de-veloped for advanced applications. They can produce large volumes of parts with maximum efficiency. Using company proprietary metal materials, one can build durable metal parts at less cost, without dependence on costly tool-ing of skilled labor.

MIT invented and patented the three-dimensional printing (3DP) process. This process is very flexible and capable of making part of complex geometry using a variety of materials. Z Corporation and Soligen use MIT's 3DP proc-ess to build 3D models. Z-Corp. offers several inexpensive models to produce functional parts. The company has been very successful in selling their prod-ucts worldwide.

Soligen, on the other hand, does not offer its machine for sale, rather it offers its services to create the mold and perform casting for customers.

EOS is a European manufacturer of laser-based rapid prototyping systems. EOS offers models of EOSINT P, EOSINT M, and EOSINT S. The primary difference between these models are build size, speed, materials, and cost. EOS's e-manufacturing using laser sintering is also explained with examples.

PROBLEMS

5.1. What are the basic principles of SLS?

5.2. Name three types of material processing capabilities of the SLS systems and briefly explain the benefits of each.

5.3. What additional peripherals and equipment is needed to run the SLS machine?

5.4. What are some applications for SLS?

5.5. Describe the three-dimensional printing process.

5.6. What are the major applications of the 3DP process?

5.7. Discuss the important components of the Z406 system.

5.8. Discuss the principles of the Z406 system.

5.9. What are some advantages/disadvantages of the Z406 system?

5.10. Briefly describe DSPC.

5.11. Name four areas in which DSPC has been used.

REFERENCES

1. 3D Systems, *www.3dsystems.com,* 2002.
2. BASTECH, Inc., *www.bastech.com.*
3. Massachusetts Institute of Technology, *http://web.mit.edu/tdp/www/whatis3dp. html.*
4. Z Corporation, *www.zcorp.com.*
5. Soligen Technologies; *www.soligen.com.*
6. EOS GmbH—Electro Optical Systems, *http://www.eos-gmbh.de/.*
7. Stotko, Christof M. *The economic effects of Mass Customization on Spare Part Business.* Vortrag auf EOS International User Meeting, 2003, Garmisch-Partenkirchen, Germany.
8. Junior, V; Stotko, C. M.; Piller, F, T. *Spare Parts on Demand (SOD) by leveraging Mass Customization manufacturing capabilities.* Konferenzband zu FISITA 2004.
9. *http://www.technologiadesign.it*
10. Technology & Design 2004: Time compression for sporting shoes with the technique of reverse engineering and rapid tooling.
11. Gallagher, T; Mitchke, Mark D.; Rogers, Matthew C. *Profitting from spare parts.* The McKinsey Quarterly, Nr. February 2005.

6

MATERIALS FOR
RAPID PROTOTYPING

The purpose of RP (rapid prototyping) is to rapidly produce physical models. Currently models can be fabricated by a limited number of materials that are suitable for testing. There is a strong need to improve the processing methods so that a wider variety of engineering materials can be produced.

The purpose of this chapter is to discuss the current RP materials and processes, provide information on a material's molecular structure, and give the advantages and disadvantages of each material. This chapter will focus on the four types of materials: polymers, ceramics, metals, and composites. The materials used specifically in liquid-based, solid-based, and powder-based rapid prototyping will also be discussed.

6.1 INTRODUCTION

Materials are the basic building blocks in RP models. The fabricated model's shape, dimensions, durability, and application are all directly related to the type of material. All industrial products are manufactured from at least one material and, most likely, many different kinds of materials. For example, an automobile contains a wide variety of materials, such as steel gears, glass windows, plastic dashboards, and rubber tires. However, in rapid prototyping the selection of materials is somewhat limited because the RP process does not allow a wide variety of materials to be fabricated. The challenge of RP is to improve the processing methods so that a wider variety of materials with "intrinsic-material" properties can be produced.

The purpose of this chapter is to provide a fundamental understanding of the different types of materials currently being used in RP. This will be observed from a nanoscopic, microscopic, and macroscopic perspective. The nanoscopic view will be examined from dimensions less than 100 nm, which means looking at molecular (atomic) bonding, crystal structures, and polymer chain length. The microscopic view will be illustrated from 100 nm to ~100 μm and includes macromolecular chain length, grain boundaries, and phase transformations (amorphous to crystalline). The macroscopic view will include dimensions greater than 100 μm, such as cracks, pits, and pores in the parts.

This chapter will cover the principles of RP materials in Sections 6.1 and 6.2. The specific materials and their properties will be covered in Sections 6.3, 6.4, and 6.5. Examples are also included in the text.

6.1.1 Nature of Materials

Manufactured materials can be divided into three broad categories: polymers (plastics), ceramics, and metals. In addition to these three categories, composites have become very important in industry as a combination of any of the three categories, whether they be metal polymers, ceramic polymers, or ceramic–metal composites. The three basic material types and their composites are shown in Figure 6.1. Composites are mixtures of the other materials and not a unique category by themselves. Semiconductor materials or ceramic–ceramic composites will not be addressed because these materials are still not used in RP fabrication.

When speaking about materials that occur in nature, it is helpful to review the periodic table of elements, shown in Figure 6.2 [1]. The periodic table classifies elements by atomic number, atomic weight, and number of electrons in their outermost shell. The atomic number and atomic weight of the elements are shown in grams per mole. Since there are 6×10^{23} atoms per mole

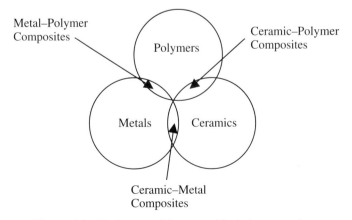

Figure 6.1 Basic material types with their composites.

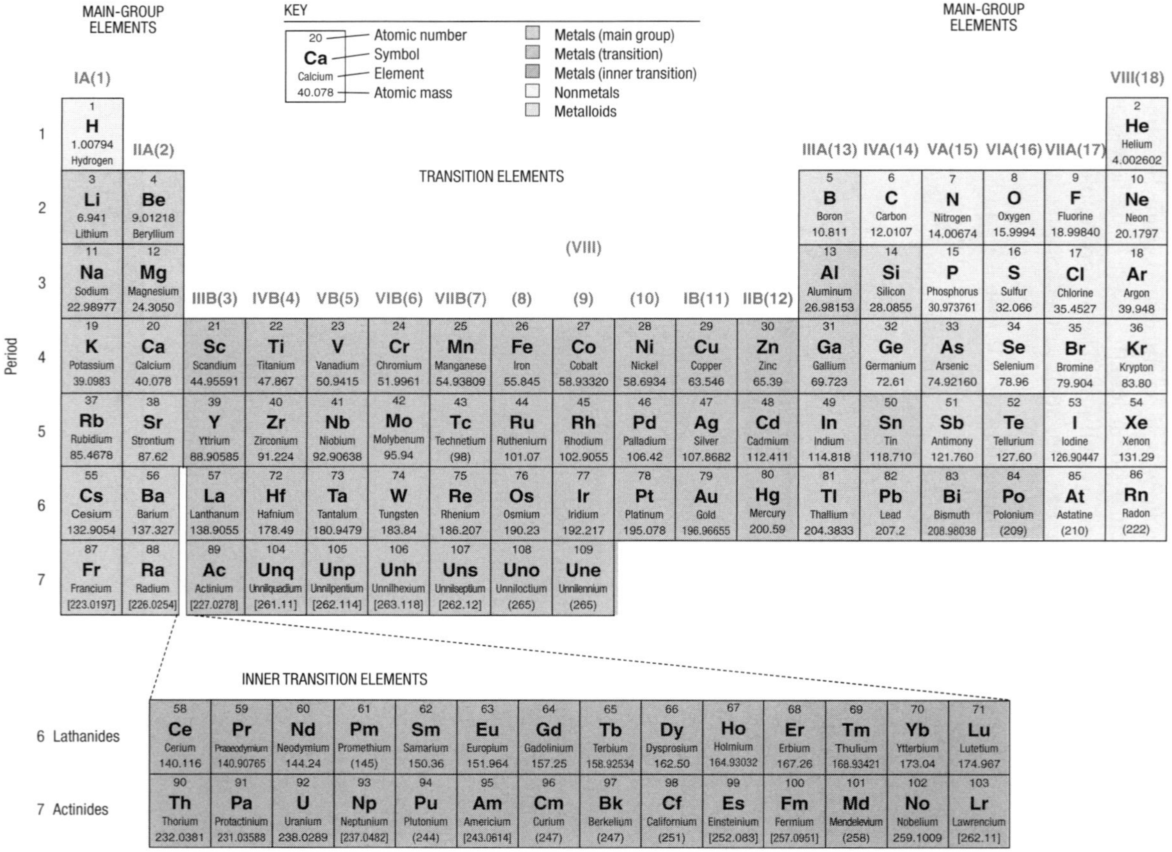

Figure 6.2 Periodic table of elements.

158

(Avogadro's number), the atomic weight of an element is the number of grams in 6×10^{23} atoms.

The nonmetallic elements, which are shown in Figure 6.2, are shaded and are typically shown in the upper right corner of the periodic table (in columns IVA to VIIA). These elements are used for forming polymers that typically contain a carbon "backbone" structure and, hence, are called nonmetallic compounds. The unshaded elements on the left and center of the periodic table are metals, while the elements that are partially shaded are semimetals. When the nonmetallic elements are combined with the metals, they form nonmetallic compounds that are called ceramics. The periodic table is also useful for determining the structure of atomic or molecular bonds.

6.1.2 Chemical Bonding and Structure

Polymers, metals, and ceramics can also be characterized by their atomic or molecular bonding—their nanoscopic features. When considering these materials, the chemical bonds that hold the atoms together in the molecule must be discussed first.

Chemical Bonding   Primary bonds are the principal type of bonding between elements; secondary bonds are lesser types of bonding such as hydrogen and van der Waals bonds. This overview will focus only on the primary bonds: covalent bond, metallic bond, and ionic bond. The covalent bond shares its electrons between atoms rather than transferring them, in its attempt to achieve a stable set of eight electrons in the element's outer shell. The organic molecule C_2H_4, ethylene, is an example of a polymer that forms a covalent bond between the C–C and C–H atom pairs. Because the electrons are tightly bound between the atoms and are not transferred between atoms, polymers have a low electrical conductivity. In addition, polymers also have a low density (~ 1 mg/m^3) because of their low atomic weight, which can be seen from the periodic table.

Metallic bonding occurs in pure metals and metal alloys. Metal atoms generally possess an ability to give up electrons because their electrons are loosely bound. Due to the overlapping energy bands in their electron clouds, metals have been characterized as ions floating in a sea of electrons, where the electrons can easily transfer from one atom to another. This accounts for their excellent electrical conductivity for which metals are so well known. Two more characteristics of the metallic bond are their good heat conductivity and good ductility. In addition, metals generally have a high density (~ 1 to 20 mg/m^3), where over 90% of the metals have atomic weights greater than 40 g/mol.

In the ionic bond, the outer electrons are transferred between metal atoms (electron donors) to nonmetallic atoms (electron acceptors) to form positively charged metal ions (cations) and negatively charged nonmetallic ions (anions). These ions are bonded together by electrostatic attraction. The reaction of magnesium and oxygen to form magnesium oxide (MgO) is an example of an ionic bond in which the Mg^{2+} cations have an ionic bond with the O^{2-} anions. These types of materials are generally classified as ceramics. Typical characteristics of ionic bonding are materials with low electrical conductivity, poor ductility, and a density that is between metals and polymers. Some ceramics have a mixture of ionic bonding and covalent bonding. Examples of this are SiO_2 and Al_2O_3.

Structure of Materials   Polymers, metals, and ceramics can also be distinguished by their crystalline or noncrystalline structures. Noncrystalline materials are also know as being amorphous. For example, when these materials are heated above their melting point, they become liquid and retain their amorphous structure even after they are cooled below their melting point. Many polymers, glasses (such as SiO_2-based ceramics), and rubbers are amorphous at room temperature. In a noncrystalline structure, the atoms are arranged randomly, without any long-range order or structure. In a crystalline structure the atoms have an ordered arrangement. Figure 6.3 shows an example of a crystalline region that is embedded inside an amorphous matrix [2].

Structural Transformations   Crystalline and amorphous structures react differently to changes in temperature, which happens to be a microscopic feature of the material. They differ especially in their phase change when they are cooled from liquid to solid state. As an example, when water is cooled below its freezing (melting) point, ice crystals form, which is accompanied by a volume change. When metals and some polymers are slowly cooled below their melting point (T_m), the atoms form an ordered crystalline arrangement, which is accompanied by a volume contraction. This volume change (per unit

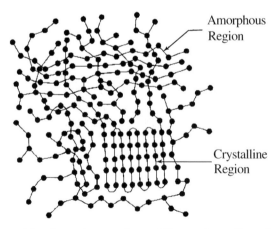

Figure 6.3   Amorphous and crystalline regions of a polymer.

weight) in response to temperature is shown by the dotted line at T_m in Figure 6.4 [2].

Amorphous materials such as glass or "highly branched" polymers, on the other hand, do not show sudden volumetric changes between the amorphous states. When an amorphous material such as glass converts from a liquid to a solid (below a certain viscosity), that point is called the *glass transition temperature* (T_g) as shown in Figure 6.4. The glass transition temperature (T_g) is very important in the production of RP tooling. When RP materials are used in tooling for plastic injection molding, it is important that the molding temperature be less than the T_g. Since an amorphous material does not crystallize into an ordered atomic structure, it does not undergo an abrupt volumetric change like the crystalline materials.

Some polymers, such as high-density polyethylene,* are almost completely crystalline (~90% by volume), as opposed to most polymers that are only slightly crystallized. Branching also affects the degree of crystallinity; for instance, a highly branched polymer cannot become highly crystalline, whereas a linear polymer can be very crystalline. The mechanical and physical properties are reflected in the degree of crystallinity of a material. As the crystallinity increases, the polymer becomes harder, stiffer, more dense, less

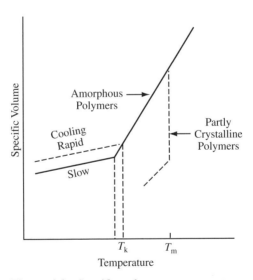

Figure 6.4 Specific volume vs. temperature.

---

*Polyethylene is a good example of different degrees of crystallinity depending on how it is manufactured. High-density polyethylene (HDPE) is highly crystalline (80 to 95%) with a density of 0.941 to 0.970 mg/m^3, and it is tougher, stronger, stiffer, and less ductile than the low-density polyethylene (LDPE), which has a lower crystallinity (60 to 70%) and a density of 0.910 to 0.925 mg/m^3.

ductile, less rubbery, and more resistant to heat and solvents. With increasing crystallinity, there is an increase in shrinkage, caused by a more efficient packing of the molecules, as was shown in Figure 6.3.

6.2   TYPES OF MATERIALS

The structure of polymers, metals, and ceramics are now going to be considered in greater detail. Examples will be taken from common materials that are used in the rapid prototyping process. More discussion will be devoted to polymers than metals and ceramics since more RP development has occurred in these materials. These materials will be examined from two perspectives: their nanoscopic and microscopic structures.

6.2.1   Polymers

Polymeric materials include plastics, rubbers, and adhesives. The term *polymers* refers to the long-chain molecules called macromolecules or giant molecules with molecular weights ranging from 10,000 to over a 1,000,000 g/mol [3]. These macromolecules exhibit strong covalent bonding between the atoms in the molecule. However, the bonding between adjacent molecules is relatively weak by 1 to 2 orders of magnitude less than the primary covalent bonds, forming van der Waals bonds or hydrogen bonds [2]. The single covalent bonds (—) between atoms such as C:C and C:H that share two electrons (:) will be referred to as C—C and C—H, respectively. Two parallel lines (==) represent a double covalent bond between atoms that share four electrons.

Mers and Polymerization   All polymers are formed of chemical linkages between relatively small molecules, called monomers, to form very large molecules, or polymers that are comprised of many monomers. A polymer is a compound formed of a repeating structural unit called a mer, whose atoms share electrons to form very large molecules. For example, the ethylene monomer molecule, C_2H_4, has one C==C double bond and four C—H bonds, as shown in Figure 6.5. Polymerization can occur during polymer processing due to heat, pressure, or a catalyst. During polymerization, the unsaturated

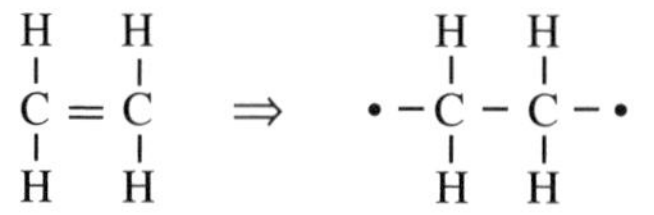

Figure 6.5   Polymerization of ethylene.

C=C double bond is broken to produce two active free electron sites (•), which can attract additional ethylene mers (repeat units) to form a poly "mer." Hence, a polymer is a chain of many mers that bond to the active sites (•).

The polymerization reaction in Figure 6.6 shows how the ethylene mers linkup to produce the polyethylene macromolecule, which is a common polymer that is produced in RP processes. If we have n ethylene mers that bond to active sites, then the polymer chain will grow in length to look like a linear molecule. Here $n = 5$ mers in the molecular chain, where n is called the degree of polymerization.

The molecular weight of a polymer depends upon its degree of polymerization as illustrated in Example 6.1.

Example 6.1

If we assume that polyethylene is produced from $n = 1000$ mers (degree of polymerization), what is the molecular weight of the macromolecule?

Solution

The molecular weight of an ethylene mer (2H—C—C—2H) can be determined from the periodic table of elements (see Fig. 6.2). The weight of each mer is $2(1) + 2(12) + 2(1) = 28$ g/mol. Since there are $n = 1000$ mers, the molecular weight of the chain is 1000×28 g/mol $= 28,000$ g/mol. This compares favorably with the molecular weight of polymer chains from 10,000 to >1,000,000 g/mol.

Chain Polymers   Commercial polymers are determined by the composition of their "backbones," the chains of linked repeating units that make up the macromolecule. Depending on their composition, industrial polymers are either carbon-chain polymers (also called vinyls) or heterochain polymers (also called noncarbon-chain, or nonvinyls). In carbon-chain polymers, the backbones are made up of linkages between C—C atoms; in heterochain polymers a number of other elements are bonded together in the backbones, including oxygen, nitrogen, sulfur, silicon, and chlorine. Both acrylates and vinyl ethers used in stereolithography, and the acrylonitrile–butadiene–styrene (ABS) used in Stratasys' FDM process [4], are all C-chain polymers. The bond lengths vary from 0.1 to 0.2 nm. Table 6.1 shows the bond lengths between typical nonmetallic atoms [5].

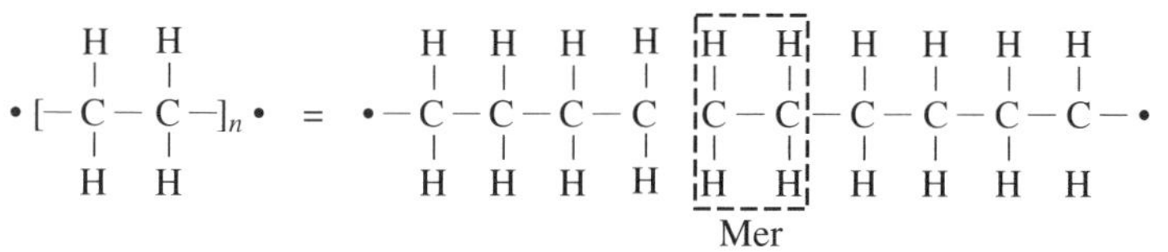

Figure 6.6   Growth of a polymer chain.

TABLE 6.1   Bond Lengths Between Atom Pairs

Bond	Length (10^{-1} nm)	Bond	Length (10^{-1} nm)
C—C	1.54	O—H	1.0
C—H	1.1	O—Si	1.8
C—N	1.5	C—Cl	1.8

Chain Length   The length of the macromolecular chains can be estimated by knowing the bond length (interatomic distance) between the backbone of the chain polymers. For example, polyethylene is a C-chain polymer, and its bond length is 0.154 nm as seen from Table 6.1. If the number of C—C bonds is known, then the average molecular chain length L in amorphous polymers is found using the equation [5]

$$L = \lambda \sqrt{m} \tag{6.1}$$

where L = length (nm)
   λ = interatomic bond length (nm)
   m = number of bonds

The bonds are not in a straight linear chain, as shown in Figure 6.6 but are actually rotated by ~110° in three dimensions. In addition, there is considerable twisting and kinking in the chains, which is shown in Figure 6.7 [5].
    Example 6.2 compares an actual chain length of a polymer with its linear-chain length.

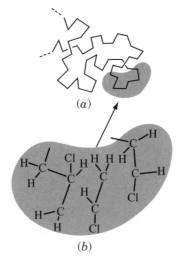

(a)

(b)

Figure 6.7   Linear macromolecule of polyvinyl chloride in liquid solution.

Example 6.2
What is the estimated length of a polyethylene molecule with 1600 C—C
bonds of interatomic bond length of 0.154 nm? Compare this with the length
of the chain if it consists of straight C—C bonds (180° apart).

Solution
Using Equation 6.1, $\lambda = 0.154$ nm for the C—C bond length from Table 6.1.
Since $m = 1600$ bonds, then $\sqrt{m} = 40$. Hence, $L = (0.154$ nm$)$ $(40) = 6.16$
nm. If straight 180° bonds are assumed, then the length of the polymer chain
would be (1600 bonds) (0.154 nm/bond) = 246 nm. The actual length of this
twisted, kinked polymer chain is only 2.5% of the length of a straight chain
of C—C bonds.

The properties of polymers depend both on the monomer type and their
molecular chain structure. Frequently, these molecular chains are in microscopic range for polymer length. There are five different chain structures that
can form linear chains, branched chains, cross-linked chains, networked
chains, and copolymer chains [2]. A schematic representation of these polymer chains is shown in Figure 6.8.

- *Linear Chains*   The polyethylene macromolecule is an example of a
  linear polymer. Linear molecules are not straight but made up of backbone atoms that are similar to spaghetti strands or major limbs of a tree.
  These polymers typically have high-impact toughness because the chains
  can slide over each other. Other examples of RP polymers that have linear
  chains include acrylics (polymethylmethacrylate), nylon (polyamide),
  and polyethylene.
- *Branched Chains*   Side branches can attach to linear-chain polymers
  during processing, similar to branches that attach to the major limb of a

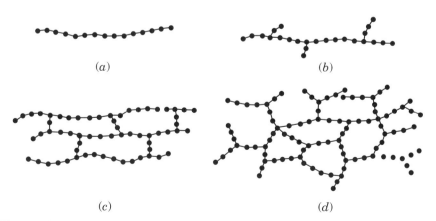

(a)   (b)

(c)   (d)

Figure 6.8   Schematic illustration of polymer chains (*a*) linear structure, (*b*) branched
structure, (*c*) cross-linked structure, and (*d*) network structure.

tree. These branches interfere with the packing efficiency and crystallinity of the linear chains. Hence the density of branched chains is lower than that of linear chains, and their strength is typically higher than linear chains. An RP example of a branched-chain polymer is ABS copolymer.

- *Cross-linked Chains* Cross-linked polymers have a 3D structure where the macromolecules are tied together by branches that are covalently bonded to adjacent chains. Cross-linking usually increases the strength, stiffness, brittleness, and hardness of the polymers. Examples include some epoxies and phenolics. The cross-linking also occurs in elastomers (rubbers) by a process called vulcanization, which is a chemical reaction at elevated temperatures with sulfur compounds. During vulcanization sulfur atoms cross-link to the various polymer chains [1].

- *Networked Chains* These are highly cross-linked 3D structures. The main difference is networked chains have many more covalent bonds that bridge the adjacent macromolecules than cross-linked chains. Networked chains typically are more rigid and have a higher strength than cross-linked polymers. Examples are epoxies, acrylates, amorphous silica (glass), and phenolics.

Cross-linked chains are stronger than linear branched chains due to their cross-linking, as discussed in Example 6.3.

Example 6.3
Explain why the strength in cross-linked structures is higher than in linear branched-chain structures.

Solution
Cross-linked structures form many more branches than linear branched structures. Also, the cross-linking is a 3D network of branches, whereas the linear branched chains have only a limited 3D orientation. The greater number of branches in 3D gives the material more isotropic mechanical properties, and it creates more resistance of the chains flowing over each other during yield when a stress is applied.

Copolymer Chains Previously, the macromolecules that were discussed had the same composition. However, when a chain is comprised of two or more types of molecules, it is called a copolymer. This arrangement of molecules in a copolymer can take several forms as shown in Figure 6.9, when two different polymers are "alloyed" together. An example of this is butadiene–styrene (BS), which is a synthetic rubber. One of the most common RP co-polymers is ABS. Here, A and S molecules form a linear AS copolymer that serves as the matrix, while B and S form a linear BS (rubber) copolymer, which acts as a filler inside the AS matrix. The combination of these two copolymers gives ABS its excellent combination of good strength and toughness [6].

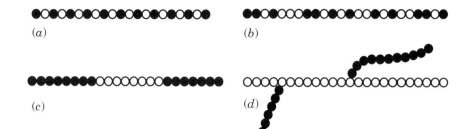

Figure 6.9 (*a*) Alternating molecules, (*b*) random molecules, (*c*) block copolymers, and (*d*) grafted copolymers. Solid circles represent one type of molecule; open circles represent another type of molecule.

Categories of Polymers   There are three categories of polymers—thermosets, thermoplastics, and elastomers [3]. These will be discussed in terms of their mechanical properties. It is important to note that this general classification presents a macroscopic view of polymers:

- *Thermosets*   A thermosetting plastic, or thermoset, is created when the chemical linkages form a rigid, cross-linked or network molecular structure. Thermosets are cured, set, or hardened into a permanent shape and usually have a high elastic modulus or stiffness that makes them more brittle. Examples of thermosets include phenolics, amino resins, acrylates, and epoxies.
- *Thermoplastics*   A thermoplastic, on the other hand, does not cure or set, can be melted when heated to a flowable state, and then rehardened by cooling. Thermoplastics form a somewhat flexible molecular structure of either linear or branched chains. Common thermoplastics include nylon, polyethylene, polystyrene, and polypropylene. Thermoplastics are generally tougher and less brittle than the thermosets but are less dimensionally and thermally stable.
- *Elastomers*   A third type of polymer is called an elastomer. As their name suggests, they exhibit a significantly high elastic strain. Elastomers include natural and synthetic rubber, silicone, neoprene, and room temperature vulcanization (RTV) rubber. Elastomers are amorphous polymers with a very low elastic modulus. They stretch when a stress is applied and return to zero strain when the stress is removed, as the molecules snap back to their original kinked configurations. The elastic modulus of elastomers is proportional to the number of cross-linked bonds of S atoms between the polymer chains, which is roughly 5% by weight sulfur [1].

The categories of polymers can be related to their material selection, as shown in Example 6.4.

Example 6.4
Which polymer category and chain structure would you select if you were designing a material for (a) tooling dies, (b) beverage containers, and (c) helmets?

Solution

(a) Tooling dies should be high strength, hard, and wear resistant. A suitable material would be epoxies or phenolics, which are thermosetting resins. A cross-linked or networked polymer would provide these properties.

(b) Beverage containers need to have ductility and impact toughness so that they will not break if dropped. A linear polymer would best serve these conditions; and polyethylene would be suitable material because it is a thermoplastic.

(c) Helmets should have a combination of high strength and high toughness to survive impacts. A good material choice would be ABS copolymer, which is a thermoplastic.

General Properties of Polymers   Polymers are used widely in industry. They are particularly useful in RP because of their low electrical and thermal conductivity and high strength-to-weight ratio. Polymers can be processed at lower temperatures because of their low glass transition temperatures. Their low density and good chemical corrosion resistance are other properties that make polymers attractive and useful in RP.

6.2.2   Metals

The material properties of metals are the easiest to determine because virtually all metals have a crystal structure. One of the most important properties is the density, which can be determined from the crystal structure. The periodic table (Fig. 6.2) provides pertinent information regarding the atomic weight of an element. Using the periodic table and Equation 6.2, the density ρ can be calculated if the crystal structure and unit cell volume are known.

$$\rho = \frac{(\text{\# atoms/unit cell})(\text{atomic mass of each atom})}{(\text{volume/unit cell})(6 \times 10^{23} \text{ atoms/g})} \tag{6.2}$$

where ρ = density (mg/m^3).

Crystal Structures and Unit Cell   Metal atoms form into a crystal structure when they solidify on cooling from the liquid state. The crystal structure determines the packing of atoms at regular and recurring positions in three dimensions. The basic geometric grouping of atoms is called the unit cell [3]. The three most common crystal structures in metals are the body-centered cubic (bcc), the face-centered cubic (fcc), and the hexagonal close-packed (hcp) structures. In these structures, it is assumed that the atoms in the unit cell are hard spherical balls. Table 6.2 shows some common RP metals with their crystalline structures and atomic radius [5].

The bcc and fcc crystal structures are shown in Figure 6.10 [5]. It is important to realize that each corner of a unit cell is shared by seven other adjacent unit cells [3].

The bcc and fcc crystal structures have unit cells of equal dimensions in length, width, and height [3]. Therefore, if the sides of the unit cell are called a, then the unit volume of the structure is a^3. In the BCC crystal structure, there are two atoms per unit cell (one atom at the center and $\frac{1}{8}$ of an atom at each of the 8 corners). There are four atoms per unit cell in the fcc crystal structure ($\frac{1}{2}$ an atom on all 6 faces and $\frac{1}{8}$ of an atom at the 8 corners). Because of the rigid arrangement of the atoms in the crystal structure, a relationship between the unit cell dimension and the atomic radius can be found. This relationship for the bcc and fcc crystal structures are shown in Equations 6.3, and 6.4, respectively:

$$a_{bcc} = \frac{4R}{\sqrt{3}} \qquad (6.3)$$

where a_{bcc} = side of unit cell (nm)
R = atomic radius (nm)

$$a_{fcc} = \frac{4R}{\sqrt{2}} \qquad (6.4)$$

where a_{fcc} = side of unit cell (nm)
R = atomic radius (nm)

Now the density of metals can be calculated. This is shown in Example 6.5.

TABLE 6.2 Some RP Metal Crystal Structures and Atomic Radii

Metal	Crystal Structure	Atomic Radius[a] (10^{-1} nm)
Al	FCC	1.431
Cu	FCC	1.278
Fe (α)	BCC	1.241
Fe (γ)	FCC	1.270
Ni	FCC	1.246
Ti (α)	HCP	1.46
Ti (β)	BCC	1.42
Zn	HCP	1.39

[a] Spherical atomic radius is assumed.

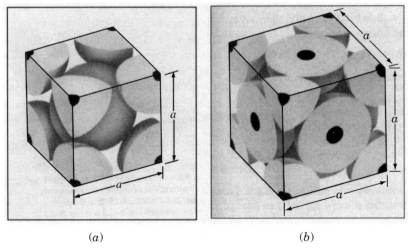

(a)   (b)

Figure 6.10   (a) Spherical ball model of (bcc) structure and (b) (fcc) structure.

Example 6.5

From the crystal structure, atomic radius, and atomic weight of copper (Cu), calculate its theoretical density and compare it to its measured density of 8.96 mg/m³.

Solution

From Table 6.2, copper has an fcc crystal structure with an atomic radius, R_{Cu} = 0.1278 nm. The density is the mass per unit volume (Equation 6.2). With 4 copper atoms per fcc unit cube and a molecular weight of 63.54 g per 6×10^{23} atoms, each unit cell has a mass of (4 atoms/unit cell) (63.54 g/6×10^{23} atoms/g) = 4.24×10^{-22} g/unit cell. The unit cell for fcc copper (a_{Cu}) can be calculated from its atomic radius (R_{Cu} = 0.1278 nm), using Equation 6.4:

$$a_{Cu} = 4R_{Cu}/\sqrt{2} = 4(0.1278 \text{ nm})/\sqrt{2}$$

$$= 0.362 \text{ nm}$$

$$\text{Density} = \text{mass of unit cell}/a_{Cu}^3 \text{ volume of unit cell}$$

$$= 4.24 \times 10^{-22} \text{ (g)}/[0.362 \text{ (nm)}]^3$$

$$= 8.94 \text{ mg/m}^3$$

Comment

The calculated density of Cu agrees fairly well with its density measured at 8.96 mg/m³, which indicates that the fcc spherical ball model is an accurate assumption of atoms in a unit cell.

Some metals undergo transformations from one crystal structure to another at certain temperatures, which are called phase transformations or allotrophic transformations. For example, Table 6.2 shows that Fe exists as two different structures (γ and α). Here an allotropic transformation occurs from fcc to bcc on cooling from an elevated temperature. Titanium also has an allotrophic transformation from hcp to bcc.

Grains and Grain Boundaries Many RP metals have randomly oriented crystal structures that are called grains. A grain is an individual crystal in which the atoms have one specific orientation. Grains usually have microscopic dimensions. When a metal is cooled down below its melting point, individual crystal or grains grow simultaneously and coalesce toward each other and randomly produce three oriented crystals as shown in Figure 6.11. Such metals are referred to as polycrystalline. Where the grains join each other, there are zones of atomic mismatch called grain boundaries. For example, Figure 6.11c shows the three grains with their associated grain boundaries [5].

Grain boundaries affect the mechanical properties of metals. For example, grain boundaries interfere with plastic deformation, and finer-grain materials generally have a greater yield strength and hardness than coarse-grain materials. Since at high temperatures grain boundaries enhance the creep rate in metals, coarse-grain materials or single crystals are preferred for lower temperatures [1].

Powder Metallurgy Polycrystalline metals can also be formed from powders through a method called powder metallurgy. Here metal powders are consolidated at elevated temperatures in inert atmospheres (to prevent oxidation of the powders). This process is called sintering, in which a high-temperature treatment causes particles to join together, thereby gradually reducing the volume of the pore spaces between them. In RP, a laser is generally used to

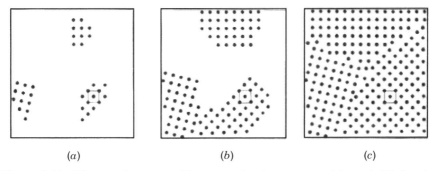

(a)   (b)   (c)

Figure 6.11 Three grains encroaching on each other as a metal is cooled below its melting point for temperatures $T_{(a)} > T_{(b)} > T_{(c)}$, where T is the temperature.

heat the powders layer by layer to replicate a CAD model, which is called selective laser sintering (SLS) [7, 8]. The RP equipment for the commercial use of SLS is manufactured by 3D Systems [9]. This type of RP process falls into the category of powder-based system, which was be discussed in Chapter 5.

Metal Alloys   In many applications, metal alloys are used rather than individual metals. Alloys are composed of two or more elements, in which at least one is a metallic element. Alloying elements are added to improve the physical properties of metals. Metal alloys fall into two general categories: ferrous and nonferrous.

Ferrous Metal Alloys   Ferrous metals are based on iron, the most common of which are steel and cast iron. There are thousands of varying types of steels; however, only a small percentage of these have been developed as RP materials. The most common ferrous alloying element is carbon, from which steel and cast iron are formed. Each alloying element enhances various properties of iron or provides new properties once the alloy is formed. Some other common alloying elements are chromium, manganese, nickel, and molybdenum.

Nonferrous Metal Alloys   Nonferrous alloys include all the other metallic elements and their alloys. Alloys are almost always more important in industry than the pure elements. Nonferrous metals include aluminum, nickel, copper, gold, magnesium, silver, tin, zinc, and titanium. The easiest to process by RP are aluminum alloys, while the most difficult to process are titanium and nickel. Aluminum alloys are widely used in the aerospace industry; therefore, the advent of more aluminum alloys for rapid prototyping would be highly beneficial.

Superalloys have become vital today due to the need for high-temperature performance in applications such as gas turbines and rocket engines. These alloys use nickel and cobalt as the base metal constituent due to their lower tendency for oxidation than ferrous alloys. Rapid prototyping is gradually entering into this new area of materials with the use of alloy powders and selective laser sintering [1].

General Properties of Metals   Metals are the most widely used materials in industry due to their many attractive properties. Although ceramics are capable of withstanding higher temperatures, they are brittle and cannot handle impacts. Polymers are light-weight, and have good strength-to-weight ratios but cannot be used in high-temperature applications. Metals, on the other hand, provide a good combination of strength and toughness and machinability. They also have good electrical and thermal conductivity and most importantly good ductility.

6.2.3   Ceramics

Definition   Ceramics are compounds that contain metallic (or semimetallic) and nonmetallic elements. Ceramics are fabricated into products through the application of heat and display such characteristic properties as hardness, strength, low electrical conductivity, and brittleness. Some common metallic elemens of ceramics are Al, Mg, Si, and Zr. Typical nonmetallic elements or anions are oxygen, nitrogen, carbon, and boron. Diamond, an allotropic form of carbon, is sometimes considered to be a ceramic because of its high hardness and brittleness. Because of their similar physical properties and similar chemical constituents, inorganic glasses such as amorphous SiO_2-based compounds are often considered to be ceramics.

Chemical Bonding   Ceramic materials such as Al_2O_3 are mixtures of ionic and covalent bonds, depending largely on the electronegativity of the elements. Small differences in electronegativity between the positively charged cations Al^{3+} and negatively charged anions O^{2-} lead to a sharing of electrons, as found in covalent bonds. Examples of the most common RP compounds include Al_2O_3 and SiO_2. The larger the electronegativity between the cations and anions, the more nearly ionic is the bonding between the cation and anion elements. Examples of such compounds are MgO and ZrO_2.

Structure of Al_2O_3   Let us examine the structures of the most commonly found ceramics: Al_2O_3. Alumina powder is used in Soligen's RP direct shell production casting. Aluminum oxide has a $2:3$ cation-to-anion ratio, where each aluminum cation has a charge of $+3$ and each oxygen anion has a charge of -2. The O^{2-} ions form a hexagonal network in an hcp crystal structure, as shown in the two fcc crystal lattices in Figure 6.12. Al^{3+} ions have a sixfold coordination around the O^{2-} ions [5]. For Al_2O_3, only two-thirds of the sixfold sites are occupied.

Structure of Silicates   Silicates are formed from the structural coordination of silicon and oxygen ions. Silicon cations (Si^{4+}) bond to four oxygen anions (O^{2-}) in a fourfold coordination, which form a SiO_4^{4-} tetrahedron, as shown in Figure 6.13a. This tetrahedral coordination is the basic structure of all silicates. Here a SiO_4^{4-} tetrahedron can polymerize with other SiO_4^{4-} tetrahedra, in a manner similar to the formation of polymer chains. For example, if $n = 2$ mers, a double tetrahedra of $Si_2O_7^{6-}$ is formed, as shown in Figure 6.13b. In this structure, four oxygen ions are connected to each silicon ion, and the center oxygen ion bridges both tetrahedral [5].

Colloidal silica is used as a binder for fabricating RP ceramic tools (Soligen) in powder-based systems. With continued polymerization (for $n = 5$ mers), a chain of five SiO_4^{4-} tetrahedra can be formed with oxygen as a bridge. Figure 6.14 shows the formation of a tetrahedral $Si_5O_{16}^{12-}$ silicate chain [5].

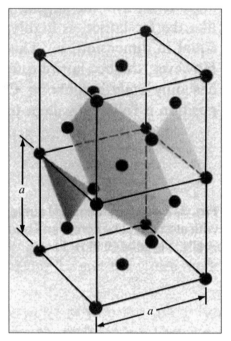

Figure 6.12 O^{2-} anion structure (shaded hexagon) and Al^{3+} cation structure (shaded triangles), where each ion has 12 close neighbors in order to preserve electrical neutrality.

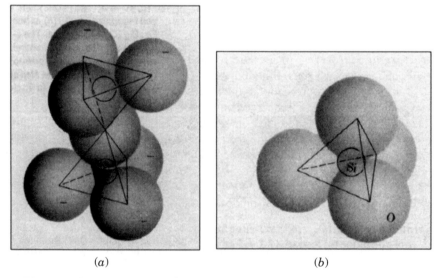

(a)   (b)

Figure 6.13 (a) Single SiO_4^{4-} tetrahedron and (b) double $Si_2O_7^{6-}$ tetrahedra.

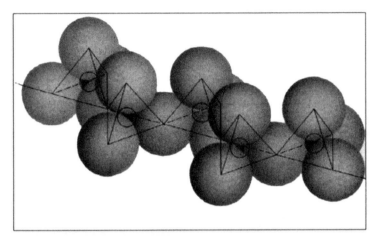

Figure 6.14   Tetrahedral silicate chain.

Just as the silica tetrahedra can polymerize into chains, three oxygen ions in each tetrahedron can be jointly shared with other tetrahedra to form a sheetlike silica structure. These tetrahedral silica sheets are shown in Figure 6.15*a*. In addition, a 3D network structure of silica can be formed that is highly cross-linked (see Fig. 6.15*b*). In this case, 3D silica is amorphous, and it is commonly called "glass" or "glassy silica."

General Properties of Ceramics   Some general characteristics of ceramics include very high strength, brittleness, and refractoriness (high melting points). Typically, ceramic materials have low electrical and thermal conduc-

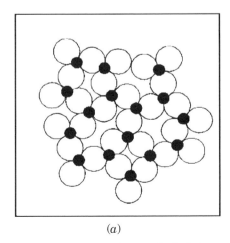

(*a*)

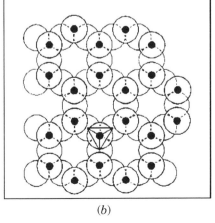

(*b*)

Figure 6.15   (*a*) Silica tetrahedral sheets shared with oxygen and (*b*) 3D network of amorphous silica.

tivity, reasonably low density, and high hardness. Ceramics are particularly useful in applications of rapid tooling, which will be discussed later.

6.2.4 Composites

Composites are nonhomogeneous mixtures of the three main types of manufactured materials: polymers, ceramics, and metals. Some metal–matrix and ceramic–matrix composites have been developed by various companies and universities for use with rapid prototyping. A composite is a material that consists of at least two phases* that are bonded together in order to obtain properties that are superior to the original components. The term composite more specifically refers to a matrix material (such as a polymer) in which fibers or particulates, such as Al_2O_3, are embedded. Embedding high modulus fibers or particulates into a lower modulus matrix improves the properties (e.g., strength, conductivity) of the host matrix. The lower modulus matrix bonds to and transfers stress to the fibers/particulates. The matrix adds greater elasticity and strain-to-failure to the composite, whereas the fibers provide enhanced strength to the composite. The reinforcing phase can improve the fracture toughness of the matrix. The fibers act as an extra barrier during extension and fracture.

Laminated Object Manufacturing   Most fibrous composites are constructed in thin layers with long fibers laid down in a single direction, and alternating that direction in successive layers so that the composite material is strong in all directions. This can be processed using laminate object manufacturing (LOM) [8]. Some composites resemble plywood with strength in only one direction, while others weave the fibers into a three-dimensional structure. The fibers may be either long, such as the tungsten–boron filaments, or shorter fibers (i.e., particles or flakes) made of glass, carbon graphite, aluminum oxide, or silicon carbide. The matrix material can be made from a high-temperature plastic such as an epoxy resin, from a metal such as aluminum, or a ceramic such as silicon nitride. The most common type of composite in use today is fiberglass, which is glass fiber-reinforced polymers.

Properties of Compositions   The advantages of composites is their high strength-to-weight and stiffness-to-weight ratio, which can be several times greater than steel or aluminum. Fatigue and toughness properties can also be greater than metals. In fact, it is possible to achieve a combination of properties not possible for any one of the original three types of materials (i.e., plastic, ceramic, or metal).

---

*A phase is here considered to be a homogeneous region, separated from other homogeneous regions by a phase boundary or a surface of discontinuity.

6.3 LIQUID-BASED MATERIALS

Liquid-based materials are thermoplastic polymers (polyamide nylons) or thermosetting polymers such as epoxies. In stereolithography, parts are built from a photosensitive polymer fluid that cures under exposure to a laser beam. This process has seen much development in the types of materials used, especially in photosensitive polymers or photopolymers. Stereolithography has generally been used on acrylates and epoxies. However, it is expected that stereolithography resin suppliers will continue to make progress in creating new materials that have selected thermoplastic properties.

6.3.1 Photopolymer Development

In the late 1960s, the first UV-curable photopolymers were developed to reduce air pollution from solvent-based coatings. Photopolymers are solidified (cured) when exposed to electromagnetic radiation with a specific wavelength including γ rays, X-rays, UV, visible light, and infrared. Radiation technology today uses electron-beam (EB) and UV curing of photopolymers as the most common commercial applications [10].

Acrylate-based photopolymers are the most widely used resin systems developed for stereolithography. Later resins were developed based on vinylether (e.g., Allied Signal's Exactomer resins) and epoxy systems. Most resins in use today are epoxies. These resin systems are developed to react to UV light in the 325-nm laser wavelength from a helium–cadmium (HeCd) laser for the SLA 250 stereolithography system (3D Systems). Other resin systems have been developed for different laser wavelengths. DSM Desotech has been in existence as an innovative manufacturer of UV/EB-curable materials, and the new division is called DSM Somos.

Perfecting the characteristics of resins has taken a long time. The early resins had very high shrink factors and low impact strengths after curing. The larger the shrink factor, the more difficult it becomes to create an accurate model [11]. For instance, an early acrylate resin, released by Ciba-Geigy in 1988, had a shrink factor of 0.8 to 1.1% and a low impact strength. The next major breakthrough in resin development was SL5143, released in 1991. SL5143 was also an acrylate resin, with a shrink factor of 0.7% and an order of magnitude higher impact strength. The increase in impact strength was vital because early models would shatter if they were accidentally dropped. SL5143 was the first resin with enough strength to allow normal handling without fear of breaking the model. The reduction in shrink factors was also very significant, and the most recent resins have 0.4% or less.

6.3.2 Photopolymer Chemistry

Definition In the RP industry, the use of polymers can be divided into two categories: (1) reactive polymers such as the photopolymers used in stereolithography apparatus (SLA) and (2) nonreactive polymers for use in the powder-based systems. In this section, we will discuss reactive polymers, while nonreactive polymers will be discussed in Section 6.4. Photopolymerization is a photochemical process that is based on photosynthesis—using light to bring about chemical change [12].

Photopolymerization can be classified according to whether each increase in relative molecular mass requires its own photochemical activation step, or whether many (thermal) polymerization steps follow the absorption of a photon. It involves the formation of cross-links between preexisting polymer chains, while photoinitiated polymerization falls into the second classification [13].

Stereolithography photopolymers were the first RP materials developed by [3D Systems, Cubital, D-MEC's solid creation system (SCS), and CMET solid object]. The process involves a photosensitive liquid polymer that cures under ultraviolet exposure from a laser. Here the polymers are typically acrylates and epoxies (thermosetting resins). The acrylates use a free-radical method for polymerization, whereas the epoxies use a cationic method of polymerization [14]. These are discussed next.

Free-Radical Polymerization   The initiation reaction is similar to the process described in Figure 6.5, in which a photoinitiator is added to the monomer fluid to speed up the rate of growth of polymer chains and overall polymerization. First, the photoinitiator (e.g., benzoyl peroxide) decomposes under laser heat to form free radicals with two reactive sites (R•) as shown in Figure 6.16.

Second, if we assume that the monomer is ethylene (only for simplicity and continuity) the R• attracts one of the C atoms in the ethylene monomer, converting the double bond into a single bond. This causes the reactive site to transfer to the other cabon atom in the mer, beginning the chain formation. This reactive step is referred to as chain initiation. Third, the mer links up with another mer (shown by dotted circle) in Figure 6.17. Then a chain reaction continues until the formation of macromolecules similar to the growing of linear chains shown in Figure 6.6. This last step is called chain propagation. Here, the mers are attracted to each other at a rate of several thousand per second, as each layer is laser cured.

The chains can be terminated in three ways: (1) when the active ends of two growing chains coalesce (called combination), (2) when the active end of one chain links up with other free radicals, or (3) when the chain removes a hydrogen atom from a second chain (called disproportionation). In the typ-

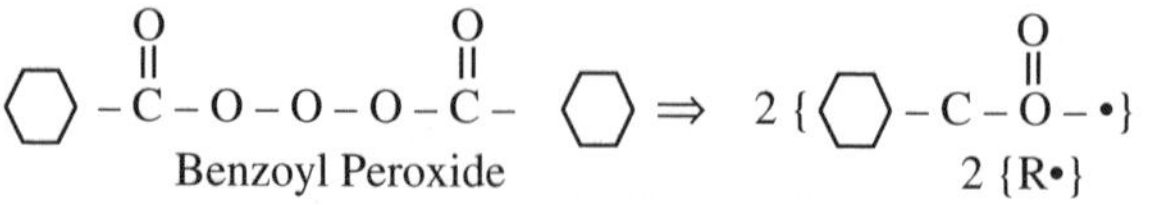

Figure 6.16   Example of free-radical polymerization.

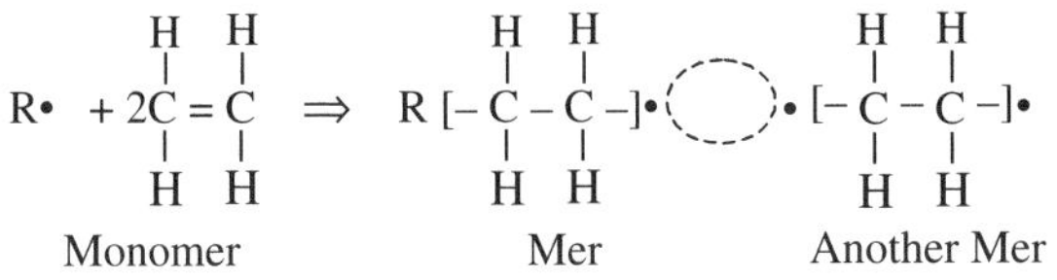

Figure 6.17 Example of chain initiation.

ical photopolymer resins, the chains usually cross-link during the curing process. Although some of the resins have good strength, not all of the properties match with the thermoplastics that they are designed to emulate [14].

Example 6.6 is used to illustrate chain reaction polymerization using free radicals.

Example 6.6
Describe the chemical reaction that occurs when polymer chains are terminated during combination.

Solution
One active site from a polymer that is capped-off with a radical chain reacts with one active site of another polymer that also has a radial on the other end. They combine and become one large chain that has all of its active sites saturated; hence the chain length cannot grow. Hence, the term chain coalescence or combination was developed.

Cationic Polymerization Cationic polymerization is a type of photopolymerization that occurs with epoxies [14]. In this case, a heat-activated photoinitiator (Q^+) cation or Lewis acid is formed. The cation breaks the O—C bond and is attracted to the active site of the oxygen atom. This creates an active site on the C atom that wants to bond an oxygen atom of another mer by breaking its O—C bond [14]. Then the chain reaction occurs and the chain length propagates. This is called an epoxy ring-opening reaction and is shown below in Figure 6.18.

Other Resins There are vinylethers that form polymer chains via cationic polymerization [14]. Over the last decade, many resins have become available in liquid-based systems. The challenges of the photopolymer chemist are not

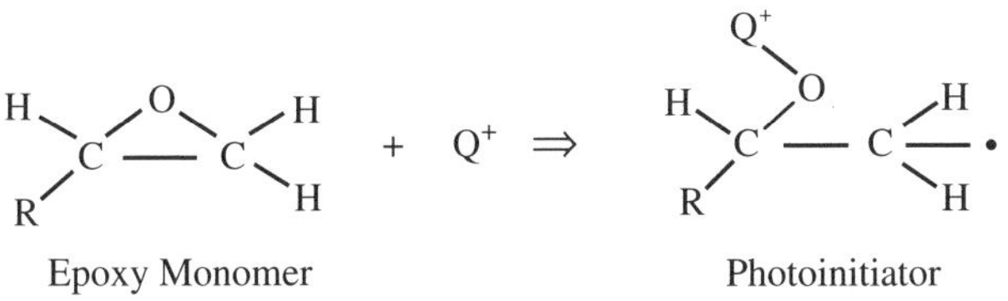

Figure 6.18 Cationic polymerization.

easy, since so many characteristics must work together to produce a usable resin. Some of the properties that are required are high photospeed, sensitivity to the proper wavelength of the lasers used on different machines, good recoating (layering) properties, high strength, low shrinkage, minimal creep of the part, high-temperature properties, and high toughness.

For fast processing speeds of large parts, medical applications from CT scanners or "rough" sketches, there is an acrylate resin RPCure 550HC that builds thick layers (0.2 to 0.5 mm) at high speed [15]. Another type of resin used for creating medical models by Zeneca is the Stereocol line for the SLA 500 and the SLA 250, which can be sterilized at medium temperatures in an autoclave. The choice of colored resin is also a useful feature for medical applications.

Elastomeric Resins Elasticity (elongation) is the property that materials have for recovering their shape after being stretched or compressed. Elastomers, such as natural and synthetic rubber, silicones, RTV materials, and polyurethane, are able to undergo large elastic deformations and then return to their original shapes when the load is removed. This is due to S atoms cross-linking the polymer chains together. In stereolithography, only the Somos resins have developed a line of elastomeric resins [8].

Typical Material Properties The physical properties for RP liquid-based photopolymer resins (acrylates and epoxies) are compared against a molded-manufactured polymer (nylon 6/6), as shown in Table 6.3. While the RP

TABLE 6.3 Properties of Materials—RP Acrylate and Epoxy Compared to Nylon 6/6

Property	Type of Material		
	Acrylate[a] RP Cure 550 HC [8]	Epoxy[b] RP Cure 200 HC [8]	Nylon 6/6[c] Molding Compound [6]
Tensile yield strength (MPa)	31	50	58
Tensile modulus (GPa)	~0.94	2.00	1.6–3.8
Elongation % at break	9%	7–16%	15–60%
Impact strength (Izod) (J/m)	Not available	Not available	29–53
Hardness (Shore D scale, unless indicated)	78	82	120 (Rockwell R)
Density at 25°C (mg/m³)	1.12	1.19	1.13–1.15

[a] Stereocol H-C, 9100 R resin.
[b] SL-5530 resin.
[c] Nylon 6/6 molding compound.

polymers approach the tensile strength of nylon 6/6, they fall short in elongation, impact strength, and hardness.

Example 6.7 will be used to predict the impact strength of acrylate.

Example 6.7
The impact strength of acrylate is not available from Table 6.3. However, based on the data in this table for acrylate and nylon 6/6, explain what you would expect the impact strength of acrylate to be.

Solution
The impact strength is a measure of the energy required to fracture a notched specimen. The energy absorbed to fracture the specimen is proportional to the area under the stress–strain for the tested material. Assuming the elongation at failure of nylon 6/6 is ~18%, the strain-to-failure for acrylate is ~$\frac{1}{2}$ that value per unit length of unfractured material (see Table 6.3). The tensile strength of acrylate is ~$\frac{1}{2}$ that of nylon 6/6. Hence, the energy absorbed is expected to be ~$\frac{1}{8}$ that of nylon 6/6. If the impact strength of nylon is 32 J/m, then we would expect the impact strength of acrylate to be anywhere from 4 to 6 J/m.

6.4   SOLID-BASED MATERIALS

Solid-based RP systems utilize solids as the primary medium to form the prototyped part [7]. Examples of these materials include the following: Sratasys' fused deposition modeling (FDM) with materials such as polycarbonate, ABS, polymethyl methacrylate thermoplastic, wax, and elastomer; Cubic Technology's laminate object manufacturing (LOM) (paper coated with a proprietary heat-activated adhesive); Solidscape's proprietary blend of thermoplastics; KIRA's selective adhesive hot pass (SAHP) paper and toner [16]; IBM's rapid prototyping system (RPS), and Solidica's ultrasonic object consolidation (UOC) [8]. Since many of these organizations have used proprietary materials, we will mainly discuss the FDM and UOC methods.

6.4.1   Polymers

Fused Deposition Modeling   Stratasys offers a unique variety of thermoplastic modeling materials for FDM systems. All are inert, nontoxic materials developed from a range of commercially available thermoplastics and waxes. The FDM process uses a spool-based filament system to feed the material into the heated zone of the machine. The material options currently include ABS called ABS (P400), a high-impact grade of ABS called ABSi (P500), investment casting wax (ICWO6), and an elastomer (E20). These are all thermoplastics that soften and liquefy in the heated zone of the machine and are deposited in 2D layers.

In 2002, Stratasys developed two other new thermoplastic materials— polycarbonate PC and polyphenylsulfone (also called polyphenylene sulfide or PPS). These two new materials have heat deflection temperatures of 125 and 207°C, respectively, which is higher than those of the previous thermoplastics [8].

The use of ABS provides the impact resistance, toughness, heat stability, chemical resistance, and the ability to perform functional tests on sample parts. As previously discussed, ABS is a carbon-chain copolymer made by dissolving butadiene–styrene copolymer in a mixture of acrylonitrile and styrene monomers and then polymerizing the monomers with free-radical initiators. Its three structural units (A, B, and S) provide a balance of properties, with the acrylonitrile providing heat resistance, the butadiene groups imparting good impact strength, and the styrene units giving the copolymer its rigidity.

Stratasys' other ABS material mentioned above is ABSi (P500). This special medical-grade ABS meets all Food and Drug Administration (FDA) Class VI requirements, that is, it can be sterilized with γ radiation and is resistant to chemicals that come in contact with medical devices. Another type of material used for FDM is elastomer (E20). This thermoplastic polyester-based elastomer has been developed for applications where toughness and durability are required for a flexible material in such applications as seals, gaskets, bushings, hoses, and tubing. Both polycarbonate and polyphenylsulfone (PPSF) were developed to enhance the strength, toughness, and temperature capability of solid-based RP thermoplastics. In addition, PPSF is resistant to gasoline, antifreeze, and sulfuric acid.

The new Genisys FDM machine has created an RP proprietary polyester (P1500), which is a thermoplastic polymer. Polyesters typically have good mechanical, electrical, and chemical properties and have good abrasion resistance. Their common applications include gears, cams, rollers, and pumps [2].

Microstructure It is important to realize that the FDM materials are deposited layer by layer and have been shown to contain interlayer porosity [17]. The open porosity is shown in Figure 6.19 for a 0°/90° automated layup. This porosity reduces the tensile properties of these materials. Hence, the mechanical properties of these materials are not as high as their injection-molded counterparts.

Specific Material Properties The properties for a wide range of thermoplastic and elastomeric materials that are fabricated by FDM are shown in Table 6.4 [8].

Example 6.8 will be used to compare the tensile strengths of epoxy and ABS.

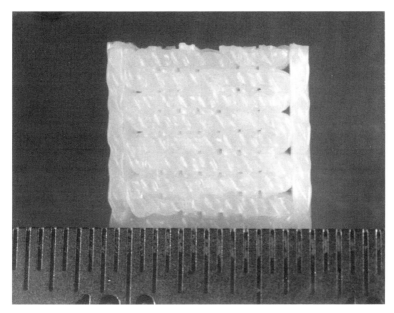

Figure 6.19   Optical photograph at 8× magnification side view of FDM tensile specimen (R6), ABS polymer. Porosity between road width and 0/90° raster layup and between 0° and 90° layers.

Example 6.8
In comparing the SLA properties for epoxy in Table 6.3 to the FDM properties for ABS in the Table 6.4, give possible explanations of why the tensile strength of epoxy is so much greater than that of ABS.

Solution
Epoxy is a thermoset, and ABS is thermoplastic. Since epoxies have more 3D cross-linking of their chains than thermoplastics, the tensile strength of epoxies are expected to be greater than that of thermoplastics. Also, the ABS fused-deposition material has microscopic porosity between the 2D layers of the parts (Figure 6.19), whereas the SLA epoxy does not have porosity. The porosity reduces the load-bearing capability of the material. Hence, the strength is lower for the FDM materials.

6.4.2   Metals

Solidica uses a process called ultrasonic object consolidation (UOC), where ultrasonic energy is used to create a solid-state bond between sheets of metal [8]. Here ultrasonic energy is imparted to the interfaces of the metal sheet. It causes frictional heating across a ~20 μm interface, where atoms can diffuse

TABLE 6.4 Properties of FDM Materials

				FDM Materials		
Property	ABS P400	ABSi P500	Elastomer E20	Polycarbonate	Polyphenylsulfone	Polyester P1500
Tensile yield strength (MPa)	34.5	37.2	6.4	63	69	19
Tensile modulus (GPa)	2.50	1.96	0.070	2.38	2.31	0.830
Elongation % at yield	>10%	>10%	>10%	N/A	7.2%	<10%
Impact strength (Izod), (J/m)	107	176	347	754	782	32
Hardness (Shore D scale)	78	76	96 (Shore A scale)	70 (Rockwell R)	122 (Rockwell R)	62

across the interface to provide a strong metallurgical bond. The typical material combinations that have been bonded together are shown in Figure 6.20.

The process is ideally suited for thin foils ~25 μm (~0.001 in.) of material. However, thicker sheets can also be used. For example, in Figure 6.21 fifteen layers of ~25-nm-thick titanium have been ultrasonically bonded together. The major advantage of this process over most RP processes is the lack of liquid-to-solid phase change in the materials involved.

This process avoids the shrinkage that occurs during the phase transformation. In addition, UOC has imbedded "smart materials," such as sensors and optical fibers, into metals.

6.4.3 Composites

Composites are mixtures of dissimilar materials. Previously, we have shown that there are polymer–matrix, metal–matrix, and ceramic–matrix composites. Here the matrix is the continuous, surrounding phase where the fibers or particulates are imbedded. The matrix usually has a lower modulus than the fibers; therefore, it transfers the load to the high-strength fibers. Composites are popular in the aerospace industry because they have high strength and are light weight. Many composites involve manual layup of 2D fiber laminates, which is very time consuming. Cubic Technologies (formerly Helisys) was the first organization to automate the layup process using laminated object manufacturing (LOM) in the RP industry [8].

	Al	Be	Cu	Ge	Au	Fe	Mg	Mo	Ni	Pd	Pt	Si	Ag	Ta	Sn	Ti	W	Zr
Al Alloys	•	•	•	•	•	•	•	•	•	•	•	•	•	•	•	•	•	•
Be Alloys		•	•		•										•			
Cu Alloys			•	•	•	•	•	•	•	•	•	•	•	•		•	•	•
Ge				•	•				•									
Au					•	•	•		•	•	•	•	•			•	•	•
Fe Alloys						•	•		•	•	•		•	•		•	•	•
Mg Alloys							•	•					•		•			
Mo alloys								•	•	•			•			•	•	•
Ni Alloys									•	•	•		•			•	•	
Pd												•	•	•				
Si													•	•				
Ag Alloys													•	•				•
Ta Alloys														•		•	•	
Sn																•		
Ti Alloys																•	•	
W Alloys																	•	
Zr Alloys																		•

Figure 6.20 Material combinations that have been joined using UOC.

Figure 6.21 Several ~25 μm (~0.001-in.) thick layers of Ti foil ultrasonically bonded.

The LOM composite material consisted of 2D laminates of paper (organic fibers) that were coated with a proprietary heat-activated adhesive (epoxy binder). The laminate is dispensed from a roll that is 114 μm (0.0045 in.) thick and is bonded by curing the epoxy binder.

Another composite is 3D Systems (formerly DTM Corporation) glass-filled (GF)/nylon composite called Duraform GF. Since this material has been processed by powder-based materials, it will be discussed in the next section.

6.5 POWDER-BASED MATERIALS

Powder-based RP systems primarily use powder as the material for building the prototyped parts [7]. In this section, powdered materials will be discussed that relate to the following RP systems: selective laser sintering of polymers and composites, direct metal deposition, and direct shell casting of ceramics.

6.5.1 Polymers

Thermoplastics Selective laser sintering (SLS) is the process that utilizes the Sinterstation (or the newer version called Vanguard) from 3D Systems. SLS is used to fabricate both polyamide (nylon), which is called DuraForm PA, and polycarbonate. These materials are laser sintered to an intermediate density and then postprocessed to increase their density.

Polymer Composites A glass-filled/nylon composite called Duraform GF is comprised of a glass-filled SLS polymer–matrix composite. It consists of glass particles imbedded inside an SLS nylon matrix. The glass-filled particles lower the tensile strength and impact strength of SLS nylon. In addition, the glass particles make the matrix more brittle than the SLS nylon. However, one advantage of this composite is its higher elastic modulus, which provides a closer match to the modulus of an actual component that is wind-tunnel tested.

Elastomers   DSM Somos 201 has been used for the SLS process. The powder is a thermoplastic elastomer that is sintered to create highly flexible parts with rubberlike characteristics. These models can be used in place of urethane, silicone, or rubber parts in such applications as moldings, gaskets, hoses, and athletic shoes. Somos 201 can withstand heat and chemical solvents and has a melting point of 156°C [8].

Powders   SLS involves heating powders with a laser to temperatures just below the melting point of the binder material (if the dispersed material is a metal). Here the temperature is high enough to sinter or bond the individual powder particles together. Then postprocessing is used to increase the density of the part.

The size of the powder particles in SLS is very critical. For example, large particles cause the surface of the part to be very course. If the particles are too fine, then the surfaces of the particles develop electrostatic charges that make the powders difficult to spread in a 2D layer. In addition, the finer particles will sinter at a faster rate than the larger particles.

Plasma discharge spheroidization (PDS) is an example of one of the new processes of producing ultrafine, spherical metal alloy powder with a controlled particle size distribution averaging 5 μm [13]. A finer particle size distribution would enable parts to be fabricated with a finer surface finish, take less time to sinter, and have higher densities. Figures 6.22 and 6.23 show that finer, spherical powders sinter to a higher density than ground powders.

Other metal powder processes utilize spherical powders that are melted by a laser. For example, the powder delivery system for Optomec's direct metal deposition process uses a spherical powder with a size range of <36 μm

Figure 6.22   Spherical powder with an average particle size of 50 μm and a sintered object created using the powder. (Courtesy of 3D Systems.)

Figure 6.23 Mechanically ground powder with an average particle size of 120 μm and a sintered object created using the powder. (Courtesy of 3D Systems.)

diameter (-325 mesh) to 150 μm (-100 mesh) [8]. Powder particles larger than 150 μm require more laser power and more time to melt. On the other hand, particle sizes finer than 5 μm are difficult to transport and require improved powder feeders and delivery systems. Fine particles also have a tendency to clog the current powder transport systems.

Selected Properties   The material properties for SLS materials are shown in Table 6.5. It is worth noting that the properties of the SLS polymers (polyamide and polycarbonate) are much less than those for the FDM of polycarbonate and polyphenylsulfone.

Example 6.9 will be used to compare the material properties between different data in the chapter.

Example 6.9
In comparing the material properties of polycarbonate that has been fabricated by FDM (Table 6.4) and SLS (Table 6.5), give an explanation of why the properties of SLS polycarbonate are much less than those of FDM polycarbonate.

Solution
The tensile strength of SLS material is less than half that of the FDM material. The modulus is about half, and the impact strength is over an order of magnitude lower. All of these properties are consistent with an SLS material that has lower density than the FDM material. In comparing Figures 6.19 and 6.22, it appears that the SLS polycarbonate has greater amounts of porosity—probably because it has not been sintered and postprocessed to full density. Certainly porosity would lower the strength, elastic modulus, and impact strength.

TABLE 6.5 Properties for Selective Laser Sintering of Powder Materials [8]

Property	DuraForm PA Polyamide	DuraForm GF Glass/Polyamide	Polycarbonate	LaserForm ST-100	Somos 201 Elastomer
			SLS Materials		
Tensile strength (MPa)	44	38.1	23	510	N/A
Elastic modulus (GPa)	1.60	5.91	1.22	137	20
Elongation at yield (%)	9%	2%	5%	10%	111%
Impact strength (notched Izod), (J/m)	214	96	53	N/A	N/A
Hardness (Shore D scale)	N/A	N/A	N/A	87 (Rockwell B)	81 (Shore A scale)

6.5.2   Metals

The main focus in RP materials is in the area of powder metallurgy (PM). Significant advances have occurred in the last few years.

Selective Laser Sintering   When dealing with the laser sintering process in the Sinterstation (or Vanguard) equipment, it is important to understand that the CO_2 laser does not sinter the metal particles together. This laser sintering is not designed to heat the metal powders to $\sim 0.5 T_{mp}$ (about half their melting point), which is required for the sintering of metal powders. Here a thermoplastic binder is used to coat the metal particles, and the binder of the powders is sintered together to form "green" parts, which are fragile. The postprocessing usually involves burning out the polymer binder and infiltrating the metal powders with a lower melting liquid metal—usually bronze or copper.

Several SLS metal powders have been used: steel, 316 stainless steel (rapid steel 2.0), and 420 stainless steel (ST-100). For example, the LaserForm ST-100 binder-coated metal powders have a particle size of ~ 36 to 76 μm. After laser sintering and binder burn-off, the stainless steel powder is 40% dense (or 60% porous). Then the steel powder is postprocessed by liquid infiltrating the powders with commercial bronze (89% copper/11% tin). The final material properties are similar to P20 tool steel, and the material is used for tooling in injection-molding machines.

SLS and Hot-Isostatic Pressing   Instead of postprocessing by liquid–metal infiltration of porous steel powders, newer processes have been developed for densifying the steel powders. One of those methods is called hot isostatic pressing (HIP), where the SLS powders are compacted to full density using high pressures (~ 100 MPa) and temperatures ($\sim 1100°$C). The HIP process has only recently been used in RP systems, but it has resulted in uniform, well-bonded grains with good mechanical properties [18]. In SLS/HIP process, the laser beam fuses the metal powder only on the part surfaces to form gas impermeable skin around the part that exceeds 92% theoretical density. (This is the fractional density at which the porosity is typically closed porosity, where it is surface connected and not interconnected with the internal porosity of the part.) The part is then evacuated inside a chamber and postprocessed to near net shape by "containerless HIP" to full density [19]. The advantage of this process over conventional HIP is that the part does not have to be placed in a container. Containerless HIP eliminates the adverse part–container reactions and the removal of the container after the HIP process has been completed. Materials such as steels, nickel-base superalloys, titanium and its alloys, refractory metals, bronze–nickel, and cermets have been processed using this method.

The hardness values of SLS/HIP material compared against conventionally processed Ti-6Al-4V alloy and containerless HIP agreed very well. Tensile specimens of this alloy showed that the "parent material" strength was exceeded, and its minimum elongation was met. Experimentation with HIP and RP was started in the mid-1990s both at Rockwell International (then Rockwell Science Center and Rocketdyne Division). Now work has continued under Rockwell Scientific. In addition a new spin-off company by the name of ODM (On-Demand Manufacturing) has been formed to produce prototype metal parts.

Direct Metal Laser Sintering Another process for consolidating metal powders to nearly full density has been developed by Electrical Optical Systems (EOS) in Germany [8]. Their process is called direct metal laser sintering (DMLS). This only uses a laser sintering process, but the DirectSteel 20-V1 powder achieves ~95% density, and no liquid metal infiltration is required. The powders used are 20 μm from materials such as steel, bronze, and nickel. The reported surface finish of the parts is very good.

Direct Metal Deposition There are three organizations that have developed a direct metal deposition process: Optomec, Precision Optical Manufacturing (POM), and AeroMet [8]. Direct metal deposition is a process that injects metal powders into a melted pool on a substrate surface as the laser scans the shape of each layer of the part. Optomec uses a process called laser engineered net shaping (LENS). Both POM and Aeromet call their process laser additive manufacturing (LAM). These processes are near-net shape and usually require a final machining to obtain a good finish. The surface finish averages ~200 to 500 μin. All the processes involve the creation of metal parts using either a Nd:YAG (neodymium:yttrium aluminum garnet) laser (Optomec), or a CO_2 laser (POM and Aeromet).

Optomec has accumulated some data on its direct metal process. In Table 6.6, the Optomec parts actually have higher strength and equivalent percent elongation as the wrought materials. In addition, the microstructure of the Optomec parts consists of a fine-grain (~3 μm) structure. POM and AeroMet have also reported an improvement in the strength of their alloys. From our limited knowledge of these materials, it can be speculated that the finer-grain structure of the Optomec materials could have been responsible for enhancing the yield strength. The finer-grain structure could have been responsible for impeding dislocation flow in the grains to cause an increase in yield strength.

A partial list of materials used in the direct metal deposition processes are 304, 316, 420 stainless steel, iron–nickel alloys, H13 and MM10 tool steels, Inconel 625, 690, and 718, titanium alloys, tungsten, Haynes 230, nickel aluminide, titanium aluminide, Mar-M 247, copper, and aluminum. Aluminum has been a desired material for RP because it is light weight and easy to machine. As shown in Figure 6.24, Optomec has had success in creating aluminum parts.

The most unique advantage of these new processes is the ability to create a controlled composition gradient of multiple materials within a single part. This has been demonstrated by both Optomec and POM. For example, a

TABLE 6.6 Material Properties: Optomec's LENS Process vs. Wrought Alloy Materials[a]

Property	LENS 316 Stainless Steel	Wrought 316 Stainless Steel	LENS Inconel 625	Wrought Inconel 625	LENS Ti-6Al-4V	Wrought Ti-6Al-4V
Yield strength (MPa)	504	245	588	406	1085	840
Ultimate strength (MPa)	805	595	945	847	1190	910
% Elongation	50	50	38	30	11	10

[a]Courtesy of Optomec Corporation.

Figure 6.24   Aluminum part as created by LENS (*left*) and machined (*right*).

gradient test sample has been processed that transitions three alloys of titanium: Ti-48Al-2Cr-2Nb, Ti-6Al-2Sn-4Zr-2Mo, and Ti-22Al-23Nb. AeroMet focuses on manufacturing titanium alloys exclusively. Optomec is also making parts using SiC-Al metal–matrix composites.

Repair of worn parts is an important aspect of direct metal deposition. Repairing expensive tools by welding destroys the heat treatment of the material, while direct metal deposition leads to a localized heat-affected zone. Direct metal deposition results in an order of magnitude less heat transfer than by welding. This new method of powder fusion results in greatly improved material properties due to the uniform grain structure over traditional welding.

6.5.3   Ceramics

Soligen's ceramic materials make direct investment casting shells (or molds) without the use of RP or wax patterns. The method, called DSPC (direct shell production castings), creates the ceramic shell layer by layer [20]. Then the layers of the ceramic shell are sintered, and molten metal is poured into the shell to produce the final part. For example, engine cylinder heads of aluminum, magnesium, iron, and steel alloys have been cast in Soligen's ceramic shells.

Aluminum Oxide   In the DSPC process, a fine layer of alumina (Al_2O_3) powder is spread by a roller mechanism on a separate building platform. Using MIT's 3D printing process, a liquid binder of colloidal silica (SiO_2) is printed onto a bed of alumina powder [8]. After sintering, the silica bonds the alumina particles together into a rigid structure. Of course, it is possible to use this technology to create a ceramic part directly, rather than using it as a shell for a casting mold.

Zirconium Oxide Zirconia is a material that is under development at Soligen, since it provides an increased cooling rate, similar to that found in sand casting. It also has good resistance to thermal shock, wear, and corrosion, as well as low thermal conductivity. Zirconia (ZrO_2) transforms from a tetragonal to a monoclinic structure when it cools down from an elevated temperature. This transformation introduces a large volume change that can initiate or propagate cracks in the part, which can cause failure. Adding CaO, MgO, or Y_2O_3 to zirconia will stabilize the cubic phase at all temperatures, thus avoiding the destructive phase transformation and creating what is called partially stabilized zirconia (PSZ). This PSZ ceramic also has better strength, toughness, and reliability than the regular unstabilized zirconia. An even newer development is transformation-toughened zirconia (TTZ), which improves on PSZ by providing improved toughness due to dispersed tough phases in the zirconia matrix. Table 6.7 shows the general ceramic properties for alumina, PSZ, and TTZ.

6.6 CASE STUDY

The following case study will be used as a design exercise for Chapter 6. Here a case is presented, and the results are expected to be studied in student groups.

Figure 6.25 shows the stress–strain data for three different polymer materials—phenolic resin, ABS, and polyethylene (PE) [2]. Suppose you were designing a component for the following applications:

- Artificial knee
- Luggage
- Door knob

Which polymer material(s) would you select for rapid prototyping? Discuss why you made your selection based upon the stress–strain response and the mechanical conditions imposed on the part. State your assumptions.

TABLE 6.7 General Ceramic Properties for Alumina, PSZ, and TTZ

Property	Alumina	PSZ	TTZ
Tensile strength (MPa)	210	455	350
Flexural strength (MPa)	560	700	805
Compressive strength (MPa)	2100	1890	1750
Elastic modulus (GPa)	392	210	203
Fracture toughness (MPa $\sqrt{m}$)	5.6	11.2	12.3
Density (mg/m³)	3.98	5.8	5.8

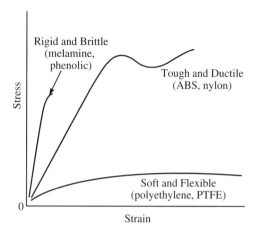

Figure 6.25 Stress–strain behavior for three polymer materials at room temperature.

Hint: Start out by listing the design requirements of the components. There may be trade-offs in the requirements, which may have to be prioritized in order to make your material selection.

6.7 SUMMARY

The rapid prototyping (RP) materials (polymers, metals, and ceramics) were reviewed from a fundamental approach—by reviewing the periodic table and chemical bonding. Polymers exhibit covalent bonding between their atoms in a molecular chain. Because metal ions can easily transfer electrons, they exhibit metallic bonding. Ceramics exhibit both ionic and covalent bonding. Composites are mixtures of the above three materials. In all cases materials can have structural transformations (e.g., amorphous to crystalline, different crystalline forms) as the temperature changes. These transformations affect the material properties.

Polymers consist mainly of carbon-chain bonds. They typically are good insulators, have either good toughness or good strength, and have a low density. Polymerization forms marcromolecules and occurs due to the breaking of C=C double bonds in the presence of heat and a catalyst. Chain reactions occur when the macromolecules form into polymer chains of high molecular weights. The chains are kinked and twisted, and they can develop branches and can cross-link with other chains. Polymers are classified into three categories—thermoplastics, thermosets, and elastomers. Examples of thermoplastics are polyethylene and polycarbonate. Epoxies and phenolics are examples of thermosets, while elastomers are an example of rubber. RP thermoplastic polymers are fabricated by fused deposition molding and selective laser sintering. RP thermosets are mainly fabricated by stereolithography.

Metals are characterized by good strength, good toughness (ductility), and good electrical and thermal conductivity. Metals crystallize into different crystal structures when they are cooled from elevated temperatures. The different crystallites are randomly oriented into grains. Polycrystalline grains can be formed from powders, and the grain boundaries can influence their mechanical properties. In many cases, alloys of different elements are used to develop the desired properties. The most common RP metals are steels, titanium alloys, aluminum alloys, and nickel-based superalloys. These metals are mainly fabricated by selective laser sintering, direct metal deposition, and ultrasonic object consolidation.

Ceramics generally are good electrical and thermal insulators. They are hard, abrasion resistant, and resist chemical attack. They exhibit both ionic and covalent bonding. Like metals, ceramics have different crystal structures. In addition, silica ceramics can be amorphous (like glass). The most common RP ceramics are Al_2O_3, SiO_2, and ZrO_2. RP ceramics are also used as direct shell castings for metals.

Composites are mixtures of the above three material categories and are mainly classified into polymer–matrix, metal–matrix, and ceramic–matrix composites. RP composites have been fabricated by selective laser sintering, ultrasonic object consolidation, and laminated object manufacturing.

PROBLEMS

6.1. How can you tell the difference between parts that are made from thermoplastic or thermoset polymer? Please explain all of your ideas.

6.2. What are the major differences between the mechanical properties of plastics and metals?

6.3. Explain the expected yield strength differences between linear, branched, and heavily cross-linked (networked) polymers?

6.4. How would the yield strength of a polymer be affected if it had a high-level crystallinity?

6.5. Explain the differences between an amorphous polymer with a glass transition temperature and a metal that is cooled down from the molten state.

6.6. Describe the design considerations of replacing a glass beverage bottle with a bottle made of plastic.

6.7. What mechanical properties do elastomers have that thermoplastics do not have?

6.8. Why do metals have a higher density than polymers?

6.9. The average distance of a polymer molecule of polyvinyl alcohol (PVA), [—C_2H_3OH—]n, is about 10 nm. (a) What *degree of polymerization* is required? (b) What is its average molecular weight?

6.10. The density of polyethylene is 1.01 mg/m^3 when it is fully crystalline. See Figure 6.17 for the dimension of the unit cell. (a) How many [—C_2H_4—] mers are there per unit cell?

6.11. For polypropylene molecules 2H—C—C—H, —R where R is CH_3, has an average molecular weight of 72,000 g/mol. (a) What is the degree of polymerization, n? (b) What is the average length of the molecule, assuming a carbon-chain polymer?

6.12. From Figure 6.10 and Table 6.2, calculate the density change for iron when Fe (γ) transforms to Fe (α). Explain what happens to the specific volume (volume per unit mass) during the Fe (γ) $\rightarrow$ Fe (α) transformation.

6.13. From the photopolymer process (in liquid-based materials), show how large polymer chains are terminated by free radicals and by disproportionation.

REFERENCES

1. W. D. Callister, Jr., *Fundamentals of Materials Science and Engineering: An Interative E Text*, Wiley, New York, 2001.

2. S. Kalpakjian and S. R. Schmid, *Manufacturing Engineering and Technology.* Prentice Hall, Upper Saddle River, NJ, 2001.

3. D. R. Askeland, *The Science and Engineering of Materials,* PWS Publishing, Boston, 1994.

4. Stratasys Inc., *http://www.stratasys.com/.*

5. L. H. Van Vlack, *A Textbook of Materials Technology,* Addison-Wesley, Philippines, 1973.

6. *Engineered Materials Handbook,* Vol. 2, *Engineering Plastics,* ASM International, Metals Park, OH, 1988.

7. C. C. Kai and L. K. Fai, *Rapid Prototyping: Principles and Applications in Manufacturing*, Wiley, New York, 1997.

8. T. T. Wohlers, *Wohlers Report 2002: Rapid Prototyping & Tooling State of the Industry Annual Worldwide Progress Report*, Wohlers Associates, Fort Collins, CO, 2002.

9. 3D Systems, *http://www.3dsystems.com/,* 2003.

10. C. Hoyle and J. Kinstle, *Radiation Curing of Polymeric Materials*, American Chemical Society, Washington, DC, 1990.

11. P. F. Jacobs, The Effect of Shrinkage Variation on Rapid Tooling Accuracy, 3D Systems North American Stereolithography User Group Proceedings, San Antonio, TX, 1998.

12. M. Hunziker and R. Leyden, *Basic Polymer Chemistry*, in P. F. Jacobs, ed., *Rapid Prototyping & Manufacturing: Fundamentals of Stereolithography,* Society of Manufacturing Engineers, Dearborn, MI, 1992.

13. C. E. Wayne and R. P. Wayne, *Photochemistry,* Oxford University Press, Oxford, 1996.

14. P. F. Jacobs, *Stereolithography and Other RP&M Technologies from Rapid Prototype to Rapid Tooling,* ASME, New York, 1996, p. 31.

15. RPCure 550HC, *Rapid Prototyping Report,* 1999, p. 7.

16. Kira Corporation, *www.kiracorp.co.jp,* 2003.

17. Optomec, *http://www.optomec.com/,* 2003.

18. D. W. Freitag, et al. Laser Directed Fabrication of Full Density Metal Articles Using Hot Isostatic Pressing. U.S. Patent 5,640,667, 1997.

19. D. Suman, M. Wohlert, J. Beaman, and D. Bourell, Processing of Titanium Net Shapes by SLS/HIP, *Materials & Design,* 115–121, 1999.

20. Soligen, *http://www.soligen.com,* 2003.

7

REVERSE ENGINEERING

Reverse engineering (RE) is the science of taking an existing physical model and reproducing its surface geometry in a three-dimensional (3D) data file on a computer-aided (CAD) system. In many cases, only the physical model of an object is available. Examples of such situations include hand-made prototypes, handcrafted items, reproduction of old engineering objects, and sculptured bodies found in medical and dental applications. In order to facilitate computer-aided manufacturing (CAM) operations of these physical models, it is essential to establish their CAD models. RE is the quickest way to get 3D data into any computer system. It is like having a low-cost but accurate X-ray machine for parts or having a 3D copier.

The principles and applications of reverse engineering by rapid prototyping will be described in this chapter. Topics such as measuring devices, model construction techniques from point clouds, data-handling and reduction methods, and their application and future trends will also be discussed.

7.1 INTRODUCTION

Reverse engineering refers to the process that creates a CAD model by acquiring the geometric data of an existing part using a 3D measuring device. Though the terminology, reverse engineering, has become more prominent in recent times, the use of RE in product development traces back a few decades. It has been widely applied in various areas such as rapid product development, casting, numerical control (NC) machining, entertainment, part inspection, medical imaging, and the like. Using RE, the development time for new products can be shortened drastically.

The reverse engineering process starts with the reconstruction of a physical part into a computerized form. Obtaining a CAD model is usually the ultimate goal of RE. The RE process aids in the interpretation of the intended design idea in order to obtain a CAD model, whereas traditional engineering turns engineering schemes and ideas into the real parts [1]. The process approaches traditional engineering from the opposite side, and this is the reason why it is called reverse engineering. It is a reversed concept of conventional engineering.

Once a CAD model is built using RE techniques, the advantages of a CAD/CAM system can be fully appreciated. The lack of an existing CAD model may be attributed to various reasons, all of which would then require the use of the RE process. These are typical uses of RE where the CAD model for an existing part is not available [2]:

- A clay model is built by a designer.
- A part has gone through many design revisions without documentation.
- The drawing of a part is lost or no longer available.

In the conventional reverse engineering process shown in Figure 7.1, the CAD model is created based on the point data sampled from the part surfaces

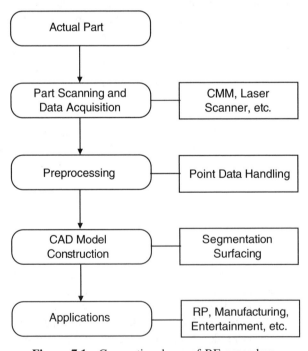

Figure 7.1   Conventional use of RE procedure.

by scanning or probing. This point data generally requires preprocessing operations such as removal of spikes and outliers and point data reduction and arrangement. For CAD model construction, the well-organized point data is segmented for a curve-based surface modeling approach or meshed when a mesh-based surfacing process is used. After fitting the point data to surfaces, the files can be saved in various CAD formats. A polygonized model such as the one with the STL format can be generated from the point data as well [3]. The computerized model plays a very important role in many application fields. In this chapter, popular 3D measurement devices and principles, 3D model generation, and applications are presented.

7.2 MEASURING DEVICES

7.2.1 Overview

Since more and more complex products are required to meet the aesthetic requirement of the customer, reverse engineering plays an important role in modern manufacturing. In the reverse engineering process, the CAD model of a product is generated from the measured point data of a physical prototype. In order to capture the shape of products, various types of measuring devices [1, 4] are used in industry, as shown in Figure 7.2. The type of

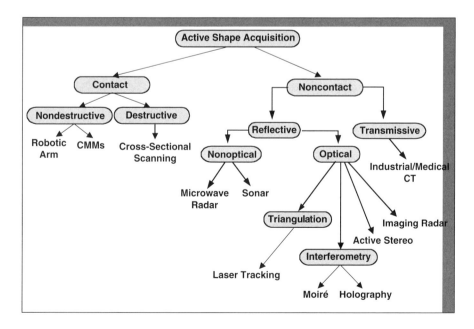

Figure 7.2 Classification of measuring devices.

measuring device can be selected by considering the part shape, required accuracy, part material, size of the part, captured data type, and so on.

Two types of measuring devices are used in sampling point data from a part surface: contact type and noncontact type. The most popular contact-type measuring device is the CMM (coordinate measuring machine), which has been used in the field extensively. This type of machine is usually NC driven and can obtain point data with an accuracy of several microns. However, it is inherently slow in acquiring point data since it needs to make physical contact with the part surface for every point that is sampled. Due to the shortcomings of CMMs, it is inefficient and difficult to measure a freeform part. Therefore, CMMs have been used for automatic inspection of simple primitive shapes such as slots, steps, holes, and pockets.

Recently, non-contact-type devices are more and more popular in industry as well as in entertainment. These devices can capture complex shaped surfaces and parts with soft materials because there is no physical contact between the product and probe. Among various noncontact-type measuring devices, the laser scanner is one of the most dominant systems. Laser scanning devices can acquire a large amount of point data in a short period of time compared to that of a contact device. Since the accuracy of the laser scanning device has greatly improved, the areas of usage are expanding greatly. In addition to laser scanners, CT (computed tomography) scanners are used for the scanning of the internal shape and material properties of parts. From the medical CT data, the human body can be visualized, and the data can be used to make implants for surgery. For industrial products the internal shape, porosity, and density information can be safely acquired using high-power industrial CT scanners.

Table 7.1 summarizes some of the major influencing factors for measuring contact and noncontact devices, the pros and cons of each, as well as their scannable shapes [5].

7.2.2 Contact-Type Measuring Devices

The CMM has been widely used in industry for a long time for inspection and reverse engineering of products. It has high accuracy, but the speed of point capturing is slower than that of laser scanners. Therefore, CMMs are mainly used for measuring parts with well-defined primitive shapes. A commercial touch probe [6] is shown in Figure 7.3. It consists of a probe head, part extension, probe, and stylus. There are many variations of touch probes in terms of the type of probe head, the type of probe, the length and type of stylus, and tip diameter.

Each point that is measured on a particular workpiece is unique to the machine's coordinate system (see Figure 7.4). The workpiece is mounted on the table of the CMM, and one can obtain the point data of the part by controlling the touch probe manually or automatically [7]. The accuracy of the captured point data depends on such variables as the structure of hardware,

TABLE 7.1   Characteristics of Contact- and Noncontact-Type Measuring Devices

	Major Influence Factors for Measuring	Pros and Cons/Scannable Shapes
Noncontact device	• Lighting condition • Surface roughness • Reflectance characteristics • Color of a part • View angle • Field of view • Depth of field • Scanning direction • Sensor fluctuations • Datum plane • System calibration • CCD resolution, etc.	Pros and Cons • Fast measuring speed • Applicable to freeform surfaces • Applicable to soft materials • Relatively low accuracy • Large number of point data Major Scannable Shapes • Complex freeform surface • Boundary edges • Casting and molding objects
Contact device	• Materials • Approach direction/speed • Probe contact point • Surface roughness • Size of details • Datum plane • Number of sampling points • System calibration • Stylus length • Accessibility • Retract distance	Pros and Cons • High accuracy • Can measure transparent part • Relatively slow measuring speed • Various application S/W available Major Scannable Shapes • Primitives • Deep holes • Interior features • Machined parts

control system, approach direction, approach speed, and atmospheric temperature. Therefore, it is important to measure a part with the proper hardware in a well-controlled environment.

Although CMMs can be used for reverse engineering, the main area of operation for CMM has been inspection. The geometric dimensions and tolerances [5] that can be measured by a CMM for inspection purposes are listed in Figure 7.5.

7.2.3   Non-Contact-Type Measuring Devices

Laser Scanning Systems   Several measuring principles are used to capture the 3D shape of a part, with the aid of vision sensors such as laser triangulation, triangulation with pattern projection, photogrammetry and stereovision, and time of flight. Among these systems, laser scanners that use an active scanning method are the most popular devices.

Generally, in laser scanning systems, the laser probe emits a laser beam to the part, and the CCD (Charged Coupled Device) cameras then capture the

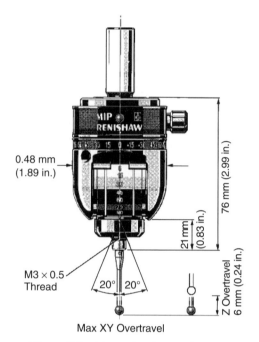

Figure 7.3 Example of a touch probe.

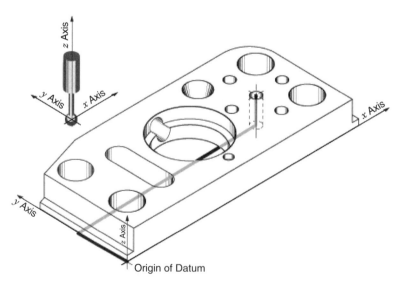

Figure 7.4 Measuring a part using CMM. (Courtesy of Brown&Sharpe.)

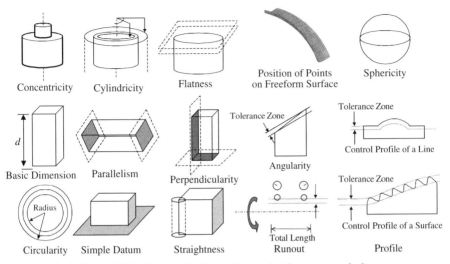

Figure 7.5   Geometric dimension and tolerance symbols.

2D image of the projected beam [8], as shown in Figure 7.6. From the captured images and the structure of the hardware system, the 3D coordinates of a part can be obtained by applying the triangulation method. A laser probe is usually mounted on a multiaxis CNC-type mechanism or on the end effector of a robotic arm.

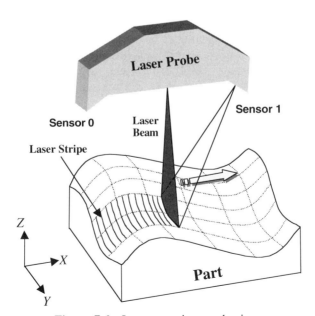

Figure 7.6   Laser scanning mechanism.

Constraints of Laser Scanning   For measuring a point on the surface of a part using a laser scanner, several constraints must be satisfied. For efficiency of explanation, some notations are introduced: P_i and N_i denote a point on the surface and the normal unit vector of that point, respectively; B_i is the bisector of the laser stripe; and L denotes the laser probe location (Fig. 7.7). All the notations above are defined on the plane where the laser beam lies. Important constraints considered for laser scanning are as follows [8]:

1. *View Angle Constraint*   The angle between the incident laser beam and the surface normal of a point being measured should be less than the limit angle γ.

$$d_i \cdot N_i \geq \cos(\gamma)$$

where $d_i = (L - P_i)/|L - P_i|$.

2. *Field of View (FOV)*   The measured point should be located within the length of a laser stripe.

$$(-d_i) \cdot B_i \geq \cos(\delta/2)$$

where δ is the FOV angle.

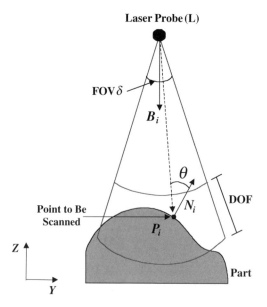

Figure 7.7   Constraints for laser scanning.

3. *Depth of Field (DOF)*   The measured point should be within a specified range of distance from the laser source.

$$l_{stand} - l_{DOF}/2 \leq \|L - P_i\| \leq l_{stand} + l_{DOF}/2$$

where l_{stand} and l_{DOF} denotes stand-off distance and DOF length.

4. The incident beam as well as the reflected beam should not interfere with the part itself.
5. The laser probe should travel along a path that is collision free.

Several other factors such as roughness and reflectance of the surface and ambient illumination also influence the accuracy of scanning results.

Types of Point Clouds   Depending on the characteristics of each measuring device, several types of captured point clouds are acquired such as scattered data, stripe data, array data, and cross-sectional data (see Fig. 7.8).

Example of Laser Scanning   As an example, Figure 7.9 shows the laser scanning of an eardrum model. Figure 7.9*a* shows a plastic model of an eardrum. The model is attached to the motorized stage and scanned using a laser scanner (Surveyor 1200 by LDI, Inc.). After scanning, the point data captured in each direction are registered on one coordinate system (Fig. 7.9*b*).

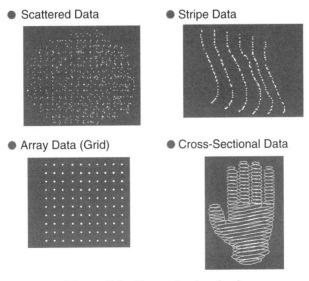

Figure 7.8   Types of point cloud.

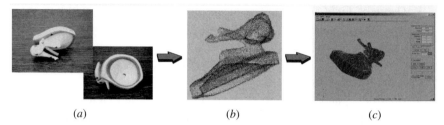

(a)   *(b)*   *(c)*

Figure 7.9 Laser scanning of an eardrum model: *(a)* eardrum model, *(b)* laser-scanned data, and *(c)* FEA mesh [9]. (Courtesy of Jaewoo, Inc.)

Finally, in Figure 7.9c, the scanned point cloud is used for the generation of CAE mesh data for acoustic analysis [9].

Computed Tomography Scanning Systems Computed tomography has proven to be an excellent nondestructive evaluation tool for measuring interior shapes, density, and porosity of products. CT scanners do not require elaborate fixturing, positioning, and part-specific programming and can generate dense, well-behaved clouds of coordinate data. Points extracted from CT images have known connectivity, surface topology, and surface normals. CT scanners have been widely used for medical applications, and in recent years, industrial CTs have been developed and utilized for manufacturing applications [10].

Medical CT Scanners In the past, medical doctors diagnosed patients using 2D X-ray images based on their experience. They imagined a 3D human body for surgery from a series of X-ray images. This made it difficult for doctors to make an accurate assessment of the patient's condition, and in the case of surgery, almost impossible to design perfectly compatible implants. By using CT scanners, doctors can visualize the patient-specific anatomical characteristics before surgery and then can generate data for machining and analysis of implants from the CT data. In a typical medical application, one-of-a-kind parts are needed for each patient, and these parts must be manufactured not by mass production but by order-adaptive production. Through a rapid prototyping process with medical-grade materials, these parts can be manufactured in a short time using CAD models from medical images.

Figure 7.10 shows the principle of spiral CT scanning, and Figure 7.11 shows a medical CT scanner [11].

Industrial CT Scanners Industrial CT has had a significant impact as a nondestructive inspection tool in aeronautical and space applications. Due to recent advances in industrial CT technology, especially in terms of data-processing methods, the field of industrial CT is expanding into several new areas of application. Data-processing methods are currently available that al-

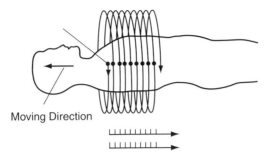

Figure 7.10   Principle of spiral CT scanning.

low for (1) CT-assisted reverse engineering, (2) CT-assisted metrology, and (3) CT-based finite element analysis. CT-assisted reverse engineering is the process of going from CT data directly to a CAD model. CT-assisted metrology is the process of extracting geometric data from a component to compare with the original design dimensions. Figure 7.12*a* shows a commercial CT scanner, and an example of a 3D model generated from a CT image is shown in Figure 7.12*b* [10].

7.3   CAD MODEL CONSTRUCTION FROM POINT CLOUDS

Two major approaches can be used to generate 3D models from point clouds: the curve based and the polygon based. However, it is necessary for point cloud data to be correctly preprocessed before building a 3D model.

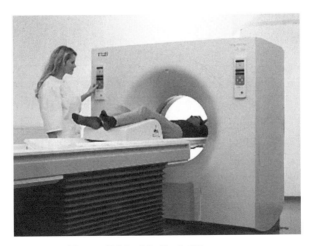

Figure 7.11   Medical CT scanner.

(a)

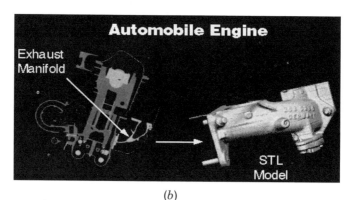

(b)

Figure 7.12 (*a*) Industrial CT scanner and (*b*) 3D model generation from CT images.

7.3.1 Preprocessing

Preprocessing operations in reverse engineering include registration, noise removal, and data reduction steps.

In the case of a complex shape, the scanner cannot capture the complete surface data from a single scan direction. Thus, the object must be scanned multiple times, with a variation in the object orientation. Every time the object is scanned, a portion of the total point cloud is obtained. Therefore, it is necessary to combine all the fragmented point clouds into one single coordinate system. This procedure is usually referred to as *registration*.

In data acquisition, noise-free point clouds are rarely found in the real world. Therefore, a *noise removal* step is important in order to create a smooth and accurate surface model often represented by a NURBS (Nonuniform rational B-spline) model.

With recent advances in scanning devices, a large number of point clouds can be gathered in a relatively short time. These highly dense point clouds, a number that could easily reach several million data points, provide a burden for further processing. Such an overwhelming amount of points not only contains a significant number of redundant points but also costs much computation time. Therefore, a *data reduction* procedure is needed to reduce the number of point clouds while maintaining the accuracy.

1. *Registration*   As mentioned before, it is necessary to integrate the point data from several views of an object into a common coordinate system, as depicted in Figure 7.13*a*. Many methods can be used for registration, all of which fall into two categorical approaches [12]: the iterative and the feature based. In an iterative approach, the ICP (iterative closest point) algorithm uses the transformation matrix calculated by repeatedly minimizing least-squares errors of the distances between the measured points and the reference CAD model. In the absence of a CAD model, the polygon model would be used as the reference model instead. This method requires the initial estimate of rough transformation by heuristics, so the accuracy of the method depends on the initial estimate. As for the feature-based approach, the several point clouds scanned in different scan directions are transformed to a single coordinate system by matching three points to three points, three spheres to three spheres, or three planes to three planes. In addition to the given feature, the geometric information of point clouds such as Gaussian curvature can be used for matching. This approach does not require the initial estimate, but it usually takes a long time to identify features. Alternatively, artificial features such as spheres (e.g., tooling balls) are attached to the part and used for matching several scanned data. Figure 7.13*b* shows the registered data of two point clouds using tooling balls.

2. *Noise Removal*   The popular noise removal methods in reverse engineering are Gaussian, average, and median methods, illustrated in Figure 7.14. All of these methods are based on statistical approximation. The average method, as the name indicates, calculates the average position of a certain set of neighboring points. In Figure 7.14*a*, three neighboring points are used, and

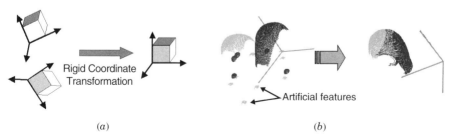

(*a*)   (*b*)

Figure 7.13   (*a*) Scheme of registration process. (*b*) Registration using artificial features.

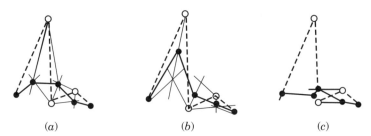

(a)   (b)   (c)

Figure 7.14 Noise removal methods: (a) average method, (b) Gaussian method, and (c) median method.

every noise-reduced point is located at the average position of three consecutive points in the raw data. The Gaussian method works similar to the average method, but with the addition of weights to the original points so that they can move along certain directions to add pulling effects. Figure 7.14b depicts the mechanism of the Gaussian method. In the median method, the noise-reduced point is located at the median of the consecutive points. Figure 7.14c shows the median method with three neighboring points. The median method works especially well when the point clouds have a certain amount of outliers. It should be noted that the above three methods work well in most cases, but these methods can cause a shrinkage of the model with a large number of iterations. Therefore, the methodology and the number of iterations should be selected with careful consideration.

3. *Data Reduction of Point Clouds* The data reduction algorithm needs to reduce the number of point clouds while keeping the sharpness of the part as illustrated in Figure 7.15. The key idea in data reduction is to

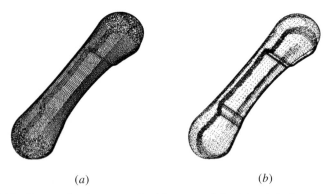

(a)   (b)

Figure 7.15 Data reduction example (phone model): (a) original point clouds with redundancy and (b) reduced point clouds.

- Remove the points on flat regions.
- Keep the points on highly sculptured regions such as boundaries, internal edges, and corners.

Various methods have been developed based on the key ideas presented above. For example, the angular deviation method reduces the amount of point clouds by comparing the angle composed of three consecutive points. If the angle is less than a specified tolerance, the midpoint is deleted. Another example is the chordal deviation method. All points in the regions that were formed by "maximum deviation" and "maximum distance" are removed. The points with a deviation larger than the maximum deviation or the points with a distance larger than the maximum distance are maintained.

7.3.2 Point Clouds to Surface Model Creation

Surface creation plays an important role in the reverse engineering process. Since the complex shaped objects are usually the targets for reverse engineering, an efficient method is crucial to create accurate surface models in a short time. Compared to the conventional surface modeling method, reverse engineering puts a greater emphasis on the creation of *freeform* surfaces. The methods of converting point clouds to freeform surfaces can be categorized into the two main approaches: curve-based modeling and polygon-based modeling, which are summarized in Table 7.2.

In the curve-based modeling shown in Figure 7.16, the scanned point clouds are first rearranged into a regular pattern, usually a series of cross sections. These point clouds are then subdivided or classified into simpler shapes of point sets. Once we have classified all subdivided point clouds, the surface fitting can be performed. Generally, surface fitting begins with curve

TABLE 7.2 Comparison of Two Modeling Approaches

Curve-Based Modeling	Polygon-Based Modeling
Segmentation, curve fitting, and skinning	Triangulation, decimation, subdivision, and triangle to NURBS fitting
Manual processes (a few weeks to a few months)	Automatic processes (a few hours to a few days)
Manual surface patch layout	Automatic surface patch layout
Class A surfaces (high quality)	Class B or class C surfaces (medium or low quality)
Surface geometry only	Surface geometry and attributes (e.g., color)
Cannot ensure continuity	Ensure G^1 continuity
Simple topology	Complex topology

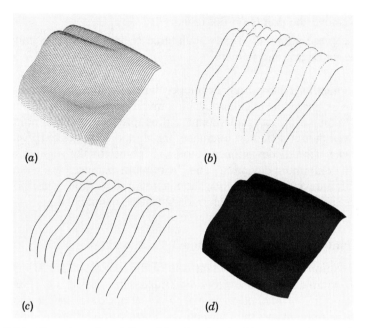

Figure 7.16   Curve based modeling: (*a*) point clouds, (*b*) cross-sectional point clouds, (*c*) fitted curves, and (*d*) skinned surface model.

fitting. The cross sections of point clouds are fitted with NURBS curves using the least-squares method. With the skinning function, the fitted curves are approximated to a NURBS surface model. Finally, each skinned surface is combined to make a valid CAD model.

Most current commercial modeling systems such as CATIA, ICEM/SURF, IMAGEWARE, SOFTIMAGE 3D, and ALIAS/WAVEFRONT can be classified into curve-based modeling systems. These systems have made it possible to create class A (high quality) surface models. However, it is difficult to create complex shaped surfaces using this approach. For instance, if an engineer wants to design a car body panel using this approach, it often takes a few weeks to months to construct a quality model.

On the other hand, the NURBS surfaces can be directly created from triangular data without fitting curves. The main characteristic of the polygon-based modeling is to construct a polyhedral model before making a NURBS model. The construction of a polyhedral model is performed by connecting neighboring points based on Delaunay triangulation, α shape, crust, or volumetric method [13–15]. The triangulated data is then refined by the decimation and subdivision algorithms. These methods have been developed recently in the field of computer graphics and would produce adequate quality data to use in application areas such as entertainment, rapid prototyping, and web visualization. If the application requires a surface model, the NURBS fitting from the triangulated data must be performed.

Some software commercially available for this polygon-based modeling includes GEOMAGIC, RAPIDFORM, and PARAFORM. Compared to the curve-based modeling systems, this procedure can construct an NURBS model in a few hours to a few days regardless of the geometric complexity of the scanned parts. Moreover, the recently developed scanning devices can capture surface attributes such as color and texture in addition to the geometry. However, it is still difficult to reconstruct fine details such as sharp corners and sharp edges with this approach.

Key Algorithms Frequently Used in Polygon-Based Modeling

1. *Decimation*   The goal of decimation is to reduce the number of triangles in the polygon model while minimizing the approximation error. Various algorithms have been developed such as edge collapse, vertex clustering, and vertex decimation. Among these, the edge collapse method would be the most popular due to its high quality of approximation. Figure 7.17 shows the basic idea of edge collapse [16]. The algorithm calculates the cost of each edge by estimating the error when it is removed. The desired number of triangles can be achieved by contracting the lowest cost edge.

2. *Subdivision*   The basic idea of subdivision is to generate a smooth and fine surface from the coarse mesh. By applying a refinement rule, we can obtain a smooth and detailed model. Subdivision surface schemes allow us to take the original polygon model and produce an approximated surface by adding vertices and subdividing existing polygons. There are many existing subdivision schemes. An example of a subdivision surface is shown in Figure 7.18. In this case, each triangle in the original mesh on the left is split into four new triangles [17].

Subdivisioning of a surface is performed very efficiently because it only uses one refinement rule. We apply just one rule and can obtain a smooth surface. The subdivisioning of a surface can be applied to various mesh types with arbitrary topology, shape, and size.

3. *Triangle to NURBS Fitting*   A series of algorithms is necessary to convert triangles to NURBS. The basic idea is to construct the initial parametric domain over the given triangles and to approximate the tensor B-spline patch with G^1 continuity. The adaptive refinement is then performed in order to

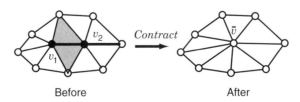

Figure 7.17   Edge collapse.

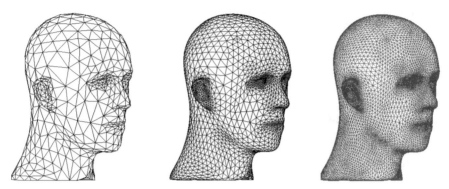

Figure 7.18   Example of subdivision surface.

achieve the required accuracy of the fitted surface. Figure 7.19 shows an example of this method.

7.3.3   Medical Data Processing

Medical images are generally represented by a voxel data set, which is constructed by an arbitrary number of slice images with the same scanning intervals. The sample points of the volume data are called the voxel (volume element) since they are spatial objects. There are two main schemes to reconstruct 3D models from volume data: *contour extraction* and *triangle extraction*. In contour extraction, each slice is processed in order to obtain contours using image processing. The extracted contours are further converted to an NURBS model by skinning. In the other method, a set of triangles are extracted between consecutive slices by the triangle extraction. This approach is useful in RP applications since no additional procedures are required to generate a file with the STL format.

Description of Voxels   Voxels can represent the geometry as well as the material properties of the scanned object. Figure 7.20*a* describes the concept of the voxel data and its geometric parameters. The parameters x_{dim}, y_{dim}, and z_{dim} represent the number of voxels in every dimension, and the parameters x_{dist}, y_{dist}, and z_{dist} describe the size of each voxel. A section of the voxel data set is usually represented by an image. The size of a cell along the x direction (x_{dist}) is equal to that of the y direction (y_{dist}) since the size of a voxel should be uniform for the entire volume data set in the 3D reconstruction process. The size in the z direction (z_{dist}) can be different since a separate parameter is used to define the distance between neighboring slices.

The voxel data values depend on the material properties of the scanned object. These data values at every grid position describe a scalar magnitude like the X-ray attenuation or the magnetic resonance within a continuous area.

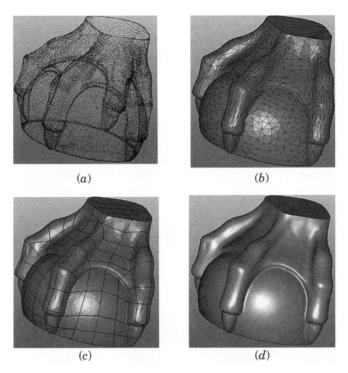

Figure 7.19 Example of conversion of triangles to a surface model: (*a*) point cloud, (*b*) triangulated and decimated data, (*c*) surface patch layout, and (*d*) NURBS surface model.

Figure 7.20*b* shows a CT image of a femur with the intensity profiles on both the vertical and horizontal lines. The higher the density of material, the higher value appears in the image. As shown in the figure, the intensity of bone is higher than that of soft tissues. The intensity value of bone ranges from 1200 to 1800 and that of soft tissue ranges from 200 to 1200 in the example.

Contour Data Extraction Various methods exist to extract contours from an image. The following steps show a standard procedure [18]:

Step 1. Apply threshold.
Step 2. Select ROI (region of interest) and remove uninterested area.
Step 3. Apply morphology operator to close gaps.
Step 4. Detect edges of object and convert them into (*x*, *y*, *z*) coordinates.
Step 5. Remove uninterested contour, if necessary.

Figure 7.21 shows an example of contour data extraction using the above procedure. The image contains not only the femur but also the acetabulum.

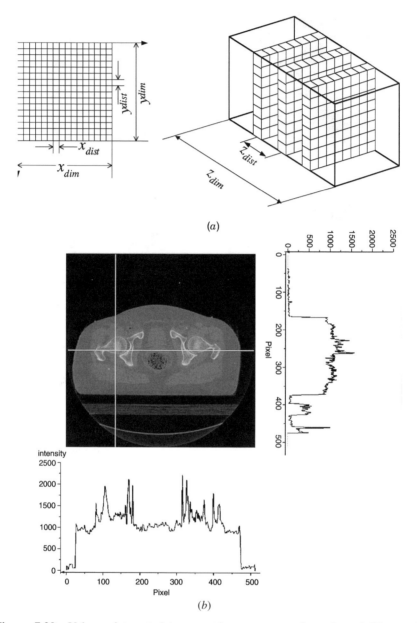

Figure 7.20 Volume data set: (*a*) geometric parameters of voxels and (*b*) material property of voxels.

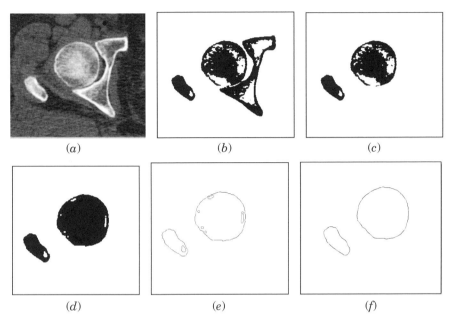

Figure 7.21 Contour data acquisition [18] (*a*) original CT image of a hip joint, (*b*) threshold image, (*c*) acetabulum removed, (*d*) apply morphology operation applied to close gaps, (*e*) extracted contours using contour tracing, and (*f*) uninterested contours removed.

Assume that our object of interest is the femur; then it is necessary to remove the acetabulum. Figure 7.21*a* is a CT image of a human hip joint with 16-bit gray level. Starting from this gray level image, a threshold operator is applied in order to separate the interested area. Figure 7.21*b* is the image using a threshold value of 1200. There exist many tiny holes and gaps in this image. After removing the uninterested area as shown in Figure 7.21*c*, morphology operators are applied to close the holes and gaps. The morphology operator includes four basic operators: dilation, erosion, opening, and closing. The dilation expands the object by converting 0-pixels, which neighbor the object into 1-pixel, whereas the erosion operation shrinks the object in the opposite way. Opening and closing operations are the combination of dilation and erosion operations. The opening is to apply erosion and then dilation. Similarly, the closing is to apply these two operators in reverse order. Although closing operation is used to fill small gaps, it is obvious that 100% accuracy will not be achieved in practice. We often need manual operations in this step. Finally, the contour-tracing algorithm is used to detect edges, and to remove uninterested contours.

Triangle Data Extraction The most popular method in triangle data generation from volume data is the *marching cube,* also known as *3D contouring* or *polygonization of a scalar field.* The marching cube is not only very simple in terms of data structure and its implementation, but also provides high-speed computation since it works almost entirely on the predefined lookup table.

The lookup table enumerates all possible topological states of a cell, given combinations of scalar values at the cell vertex. Since one grid cell is defined by 8 vertices and scalar values at each vertex, the number of topology state is equal to 2^8 (256) cases. These 256 possible configurations of triangle can be reduced to 15 cases due to the symmetry such as rotation and mirroring. Figure 7.22 shows the 15 cases for isosurface generation [19]. A dark vertex in the figure indicates the vertex value is greater than the prespecified value.

For each grid cell, the marching cube algorithm tries to create a set of planar triangles that best represents the isosurface of that grid cell. If one or some of the values at the vertices of the cell are less than the user-specified isovalue, this cell should contribute to construct an isosurface. Each cell is treated independent of the other in order to construct whole isosurfaces of an entire volume.

Figure 7.23 shows an example of isosurface generation in a single cell. If the value of the dark vertex v_1 is greater than the given prespecified isovalue, the topological state of this configuration would belong to the case 1 in the lookup table shown in Figure 7.22. A triangular facet would be created whose three vertices are located between edges e_0, e_1, and e_2, respectively. The exact

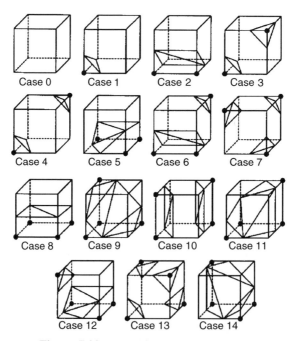

Figure 7.22 Marching cube lookup table.

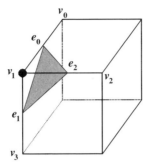

Figure 7.23   Example of isosurface generation.

positions of the extracted triangle vertices are then determined by calculating the linear interpolation of their vertex value. For example, in Figure 7.23, if f_0 and f_1 are the vertex value of v_0 and v_1, respectively, then the position of the triangle vertex p is determined by the equation

$$p = v_0 + \frac{\text{isovalue} - f_0}{f_1 - f_0} v_1 - v_0 \tag{7.1}$$

7.4   DATA-HANDLING AND REDUCTION METHODS

Data handling and reduction have become an important issue in reverse engineering. Many data-handling and reduction methods have been proposed in the area of image processing, but they were mostly designed for dealing with the meshed point data. Only a few methods were developed that could be directly applied to the point data generated from measurement devices. In this chapter, 2D and 3D grid-based point data reduction methods are introduced in order to apply to a three-dimensional point cloud data.

7.4.1   Uniform-Grid Method

The amount of point data can be reduced by dividing them into grids and by sampling a representative point from each grid using a uniform-grid method as described below [20]. An array of grids perpendicular to the scanning direction (z direction) is used for extracting points from the point cloud data. Since z values, as discussed in the previous section, are more prone to errors due to the characteristics of laser scanners, the median filtering is used with the grids. First, a grid plane consisting of equivalent-sized grids perpendicular to the scanning direction is created. The data reduction ratio is determined by the size of a grid that can be user defined. The smaller the size of the grids, a greater number of points is sampled from the entire point cloud. After creating a uniform-grid plane, all the points are projected on the grid plane and each grid is assigned with the corresponding points. Then, one point from each grid is selected based on the median-filtering rule. In fact, the points within each grid are sorted with respect to the distances from the grid plane, and a point that is located in the middle is chosen as shown in Figure 7.24. When the number of points within a grid is n, the $(n + 1)/2$ point is selected if n is odd, and the $n/2$ point or the $(n + 2)/2$ point is selected if n is even.

Using median filtering with uniform grids, those points that are regarded as noise are likely to be discarded. This method shows better performance if the scanned surface is perpendicular to the scanning direction. Furthermore, this method is good for maintaining the original point data set as it selects points rather than changing their positions. The uniform-grid method is especially useful in cases where data reduction needs to be done very quickly for parts with relatively simple surfaces (see Fig. 7.25).

7.4.2 Nonuniform-Grid Method

In the case of applying the uniform-grid method, some points for which the part shape drastically changes, such as edges, can be lost because no consideration of part shape is provided. In reverse engineering, it is critical to accurately recreate part shape, and the uniform-grid method has limitations in this regard. In this section, nonuniform-grid methods in which the size of the grids can be varied based on the part shape are introduced [21]. Two levels of nonuniform-grid methods are described: one directional and bidirectional. They can be applied considering the characteristics of the measured data.

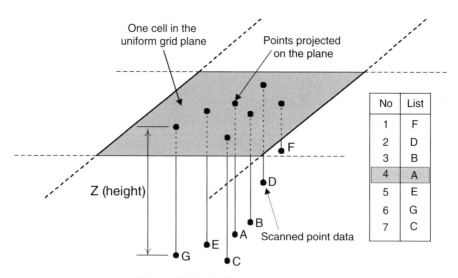

Figure 7.24 Uniform-grid method.

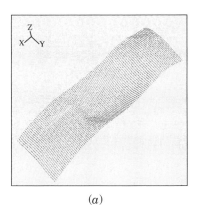

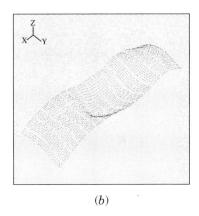

(a)   (b)

Figure 7.25 Result of uniform-grid method for freeform shape: (a) initial data and (b) reduced data.

When a part consisting of simple surfaces has to be measured by using a stripe-type laser scanner, it need not be scanned densely in all directions. In the case that the point data has more points along one direction (v direction) compared to the other direction (u direction), the one-directional nonuniform-grid method is appropriate enough to capture the shape of the surface and is, therefore, recommended. On the other hand, when a part to be measured has complex and freeform surfaces, the point data is supposed to be dense along both the u and v directions. In this case, the bidirectional nonuniform-grid method is more appropriate than the one-directional nonuniform-grid method.

One-Directional Nonuniform-Grid Method In the one-directional nonuniform-grid method, points are sampled from the point cloud using an angular deviation method. The angular deviation method selects the points based on the angle that is calculated by the vectors created from three consecutive points. The angles represent the curvature information, so when the angle is small the curvature is small and vice versa. Using these angles, the points with high curvature can be extracted. The size of the grids along the u direction is fixed by the interval of laser stripes, which has already been determined by the user. Regarding the v direction, the size of the grids is determined depending on the geometric information of the part shape. The points extracted by angular deviation represent high curvature areas, and they need to be preserved during data reduction in order to accurately express the part shape. Thus, after extracting points by using the angular deviation method, the grid along the v direction is divided based on the extracted points as shown in Figure 7.26a. When dividing the grids, if a grid is larger than the predetermined maximum size, it is divided not to exceed the maximum grid size as shown in Figure 7.26b. Then, median filtering is applied to the points in each grid. This will result in a representative point for each grid as

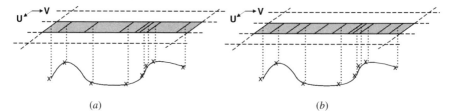

Figure 7.26   One-directional nonuniform-grid generation.

in the case of the uniform grid method. The final retained points in this method include the points selected from using median filtering for each grid and the points extracted by angular deviation. Using this method reduces the point data more effectively while maintaining the accuracy of the part shape, when compared to the uniform-grid method.

Bidirectional Nonuniform-Grid Method   In the bidirectional nonuniform-grid method, the normal vectors of individual points are calculated and data reduction is performed based on this information. First, the point data is polygonized using triangles. The normal vector of a point is calculated by averaging all the normal vectors of neighboring triangles.

Upon calculating the normal vectors of all the points, a grid plane is generated. The size of the grid is predefined by the user, which depends on the intended data reduction ratio for the given part shape. If the point data needs to be reduced greatly, the size of a grid will be increased, whereas if the point data needs less reduction, the grid size will be decreased. By projecting the points on the grid plane, the points corresponding to each grid are grouped and the normal values of these points are averaged. As a criterion for the subdivision of grids, the standard deviation of the point normal values is used. The level of standard deviation is predetermined considering both the part shape and the desired point data reduction ratio. If the standard deviation in a grid is, for instance, large, it indicates that the part geometry corresponding to the grid is complicated, and therefore, further subdivision of the grid is required in order to sample more points. The process of subdividing is shown in Figure 7.27. If the standard deviation of a grid is larger than the given value, the grid is subdivided into four cells. This process repeats until the standard deviation of a grid is smaller than the given value or the grid size reaches the minimum limit specified by the user. The minimum size of a grid varies depending on the complexity of the part shape. Upon completion of grid making, a representative point is selected from the points belonging to each grid using median filtering. Figure 7.28 illustrates an example of bidirectional nonuniform-grid method. This bidirectional method extracts more points compared to the one-directional method, thereby representing the part shape more accurately (see Fig. 7.29).

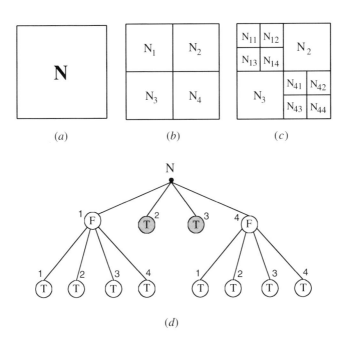

Figure 7.27   Nonuniform-grid method: (*a*) initial cell, (*b*) first split, (*c*) second split, and (*d*) quadtree structure.

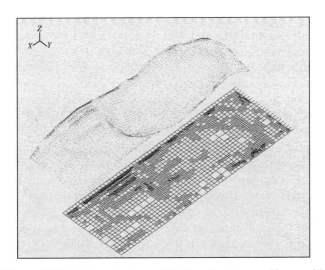

Figure 7.28   Reduced data and bidirectional nonuniform grid.

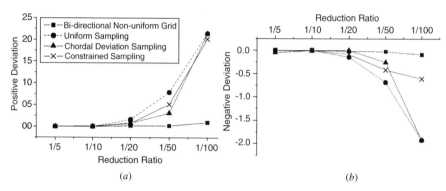

Figure 7.29 Error analysis of bidirectional nonuniform-grid method: (*a*) positive deviation and (*b*) negative deviation.

7.4.3 Three-Dimensional Grid Methods

The 3D grid method can handle the entire surface of a 3D object whether it is a single point cloud or multiple point clouds [22]. In the case of multiple point clouds for a single object, they need to be registered under one coordinate system. The proposed method uses the normal values of points on the part surface, from which 3D nonuniform grids are generated using standard deviation of normal values. Data reduction is performed by selecting one representative point and discarding the other points from each grid.

Normal Estimation   To obtain geometry information of a part surface, various curvature values such as Gaussian, mean, and principal curvatures have been used. Since it is hard to extract geometry information of a part surface directly from point data, the model is usually triangulated; thereby, curvature data or normal values of triangles are obtained. In the proposed method, the normal values of points, so-called point normals, are used instead to extract geometry information.

Depending on scanning devices or scanning methods, point data can be classified as structured or unstructured. The point data obtained by laser scanners or Moiré scanners belongs to the structured category. The point data obtained by portable CMMs or hand-held scanners belongs to the unstructured. In estimating point normals, Delaunay triangulation is usually used regardless of point data types. For unstructured point data, point normals are calculated after Delaunay triangulation. But for structured point data, different normal estimation methods can be used depending on the structures.

In the case of point data generated by a laser scanner with a stripe-type light source, the point data has an inherent order. Since a scan path is defined as a series of line segments, each line in the path is ordered as well as the points in each scan line. The normal estimation values can be calculated quickly using this pattern.

In the scanning operation of noncontact devices such as laser and Moiré scanners, an object surface needs to be measured from one direction or from multiple directions. When a single direction is used, the part surface can be scanned at once with one orientation. But for the one with multiple directions, it needs to be scanned several times while changing its orientation. In general, an object should be scanned with various orientations if the entire surface of a 3D object is needed. The 2D Delaunay triangulation can be used for normal estimation of the point data obtained from a single direction. After triangulation, point normals are calculated from circumference triangles.

Three-Dimensional Grid Subdivision Using an Octree   Spatial decomposition methods based on octree structures have been proposed for use as approximate representations of geometric objects [23, 24]. The basic concept of the octree representation consists of placing the object of interest in a parallelepiped, typically a cube, which totally encloses it. As shown in Figure 7.30, this parallelepiped is then subdivided into its eight octants, which are then recursively subdivided a number of times based on the criteria defined by application. In approximating a geometric object, the octants completely inside or outside the object are not subdivided further while those octants that contain a portion of the object's boundary continue to be subdivided to the required level. The concept of octrees is used here for data reduction. The criterion used for subdividing a cube is the standard deviation of point normal values.

Grid Generation and Subdivision   Upon completion of calculating the point normals, the normal values are stored using a point data structure, which has x, y, and z coordinates and x, y, and z normal components. Then, all the point clouds belonging to an object need to go through a registration process.

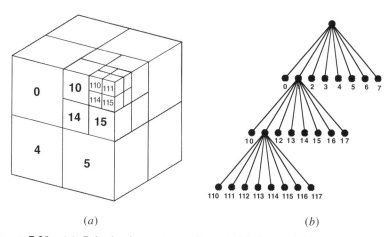

(a)   (b)

Figure 7.30   (a) Cube having octants after subdividing and (b) octree structure.

After the registration is completed, a bounding box is created (see Fig. 7.31a). The shortest axis of the bounding box among x, y, and z axis is selected to decide the number of initial grids. It is then divided by a user-defined number to make an initial grid into a cube (see Fig. 7.31b).

Among these grids, unnecessary ones that do not contain any points are eliminated. Then each grid is subdivided using octree decomposition, and empty grids are again eliminated. For subdivision, the standard deviation of point normals within each initial grid is calculated. When the standard deviation is larger than the user-defined tolerance, a grid is divided into eight grids. This subdivision process continues until the divided grids meet termination conditions. The termination conditions are met when the standard deviation of point normals within a grid is smaller than the given tolerance or a grid contains only one point (see Fig. 7.31c).

Extraction of Points   As a result of grid subdivision, many grids are generated where the part geometry drastically changes, whereas a few grids appear where the part geometry shows little change. From these grids, the points

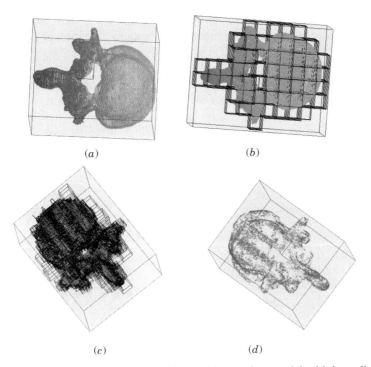

(a) (b)

(c) (d)

Figure 7.31   Example of a human backbone: (a) complete model with bounding box, (b) initial grids, (c) nonuniform 3D grids, and (d) reduced point data.

that can represent part geometry are extracted. In selecting a point that represents the points within a grid, an average of normal values is used. Therefore, a point whose normal value is closest to the average of the points within a grid is chosen. The selected point is regarded as the most representative point among the points within the grid (see Fig. 7.31*d*).

In this method, the level of data reduction is basically determined by two factors: the number of initial grids and the size of the user-defined tolerance. For surface fitting of a plane as an example, the data points need to appear within a certain interval so that a surface fitting operation can be performed. In this case, the size of the initial grids is determined by these intervals and the number of initial grids is decided accordingly.

In the developed program, the number of initial grids is determined before the tolerance. The tolerance is the main factor in data reduction; the smaller the tolerance the more points are left, and vice versa. If the user wants the remaining points to be distributed in certain grids to keep the accuracy, the smaller size initial grids must be used. If the user wants less number of points while keeping the accuracy, the reduction process should start with a lesser number of initial grids with a high tolerance.

Analysis   A phone model shown in Figure 7.32*a* was used for evaluating the performance of the 3D grid method. In this case, the point data was simulated by converting a surface model into an STL model with a small tolerance. The nodes of the STL model were regarded as measured points and no noise was added to the point data so that they can be considered as point data that already had noise filtering.

Figures 7.32*b* through 7.32*e* show the phone model data for different methods using data reduction of 90%. The phone model shown in Figure 7.32*b* by the 3D grid method shows more points distributed at the edges compared to those shown in Figures 7.32*c* and 7.32*d* by uniform or space sampling methods. For the point data reduced by chordal deviation sampling, the edges seem to be well preserved as shown in Figure 7.32*e*, however, the remaining point data does not perform as well for surface fitting. When using the same number of points, the phone model generated by the 3D grid method keeps better details at edges.

To compare the size of errors between the point clouds generated by different reduction methods, each point cloud was segmented, and then each segmented point cloud was fitted to a surface using a commercial RE software. The difference between the reference surface patch and the surface fitted by the point data for different reduction methods is illustrated in Figure 7.32. The maximum error of each method is shown in Figure 7.33*a* while Figure 7.33*b* shows the average error. These graphs show that the reduced point data by the 3D grid method keeps better accuracy than those by the other methods. As the reduction ratio increases, the error increases drastically in the other methods.

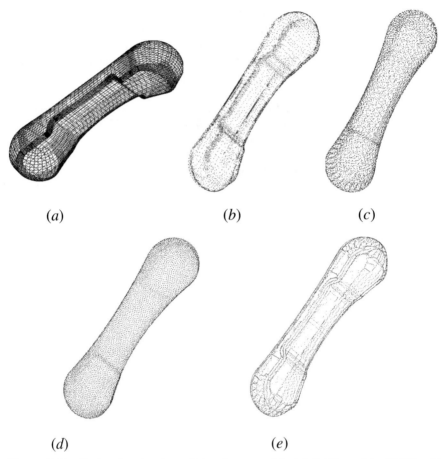

Figure 7.32 Reduced point data of the phone model: (*a*) CAD model, (*b*) 3D grid method, (*c*) uniform sampling, (*d*) space sampling, and (*e*) chordal deviation sampling.

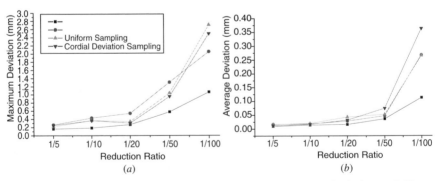

Figure 7.33 Error analysis of 3D grid method: (*a*) maximum deviation and (*b*) average deviation.

7.5 APPLICATIONS AND TRENDS

7.5.1 Applications

Since RE is the method of creating a CAD model from a physical part, RE can be applicable to any field that needs to obtain design data from a real object. Among the various applications, we present three significant application areas: manufacturing, entertainment, and medical engineering. Application details of each area are described in the following sections.

Manufacturing   Manufacturing is one of the most important areas where RE has been used since the beginning. Styling, mold design, customer product design, and inspection are the primary applications in manufacturing.

Reverse engineering is well known as the key process in designing a car body [25]. An aesthetic part such as a car body is very difficult to draw directly by mathematical formulas in 2D or 3D. The alternative solution is to create a CAD model from a clay model of the car body. Figure 7.34 shows the measured data and the CAD model created from the data. Similarly, CAD model generation from a mockup is one of the major processes in any styling job.

When creating a mold, manual modifications occur frequently, and the final shape of the mold often changes from the original design data. Therefore, the original CAD model is not valid any more, and a new CAD model for the modified mold must be obtained. The following figures illustrate an example of reconstructing a CAD model from a mold. Figure 7.35*a* shows the mold of a winding frame that is used for making a deflection yoke coil included in monitors. A reconstructed CAD model of the winding frame is shown in Figure 7.35*b*.

Inspection is the process that compares a manufactured part with its original CAD model to verify whether the part coincides with the original design data. The RE process can be directly applied to inspection. In inspection,

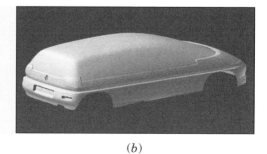

(*a*)	(*b*)

Figure 7.34   Design of car body: (*a*) measured data and (*b*) generated CAD model of car body. (Courtesy of Capture3D, Inc.)

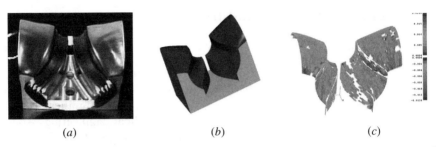

(a)   (b)   (c)

Figure 7.35   RE application in mold design: (*a*) winding frame, (*b*) reconstructed CAD, and (*c*) inspection.

either the measured point data are directly compared with the CAD model or the created surface model is compared with the CAD model. Figure 7.35*c* shows an example of inspection of the mold of the winding frame.

Customization of products is one of the key trends in present markets. For example, consumers prefer garment products suited to his or her body over ready-made products. Customization is also an important factor in RP applications. Clothes, helmets, and shoes are typical products that can benefit from customization. In order to develop products fitted to a human body, the body must be scanned and its CAD model needs to be created. Figure 7.36 illustrates some examples of customized products.

Entertainment   The desire for entertainment is rapidly increasing. Movies, games, and toys occupy a large part of children's and adults' lives. For some movies these days, animation and computer graphics often play an important role. Since human and animal animation models are very complicated, they are created using reverse engineering techniques in many cases. Avatars used in 3D games are similar to the ones in the movies. Similarly, in the toy industry, mockups of characters are created first and then converted to CAD models. As observed, the parts that consist of freeform and complicated sur-

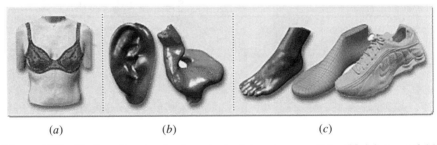

(a)   (b)   (c)

Figure 7.36   Design of customized products: (*a*) underwear, (*b*) artificial ear, and (*c*) shoes. (Courtesy of INUS Technology Inc. [26])

faces are targets of the RE process. Applications of RE in the entertainment industries are illustrated in Figure 7.37. It shows an animation character from *Geri's Game,* which received the Academy Award for Best Animated Short Film in 1997 [27]. The control mesh of the human character was created by digitizing a full-scale model sculpted out of clay. The smooth (polygon) surfaces were created by subdivision techniques.

Medical Engineering   Medical engineering is one of the newer areas where rapid prototyping techniques can be successfully applied. Since human bones have complicated shapes, NC machining of these parts is difficult. RP techniques can be used to fabricate them, with the CAD model being created by RE techniques. Measurement data for human bodies are acquired by CT/MRI scanners or noncontact measuring devices such as laser scanners. Generated CAD models of bones can be used for many applications such as implant design and surgery planning. Figure 7.38 shows CAD models of knee bones, skull, and teeth reconstructed by the RE process [28].

7.5.2 Future Trends

The ultimate goal of RE is to fully automate the process from measurement to CAD model generation. The focus of hardware development will be on measuring systems that can capture accurate data for a wide area in a short period of time. Along with the measurement hardware, measurement planning software should be developed to facilitate automated measuring. Several polygon-based RE software programs have greatly reduced the time required for creating a CAD model. But in order to create A-class surfaces, much time and effort is still needed.

Until now, the RE process has mainly dealt with the reconstruction of the outside shape of a part. But there exist many cases where the internal shape

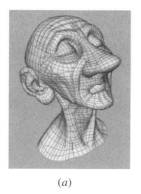

(*a*)   (*b*)

Figure 7.37   Examples in animation (*Geri's Game*): (*a*) control mesh model and (*b*) model with rendering.

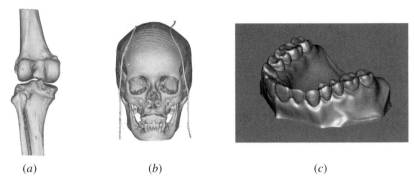

(a)   (b)   (c)

Figure 7.38 Medical application of RE: (a) knee bones, (b) skull, and (c) teeth. (Courtesy of Geomagic, Inc.)

of the part is important. It is expected that the technology for industrial CT scanners will advance and research/applications for internal shape reconstruction will become more available. Currently, CT has been applied to medical image acquisition and partially to industrial parts, but applications are expected to be extended to most industries.

Another trend that we can notice is the development of polygon-based software. Due to the simplicity of representation of polygon models, its usage has become popular. For instance, more CAM systems dealing with polygon models are under development.

In addition to the applications described above, many other potential areas of application exist. Wherever digital models for real parts are required, the RE technique can be applied. Some example areas include cultural assets, archeology, web3D, and virtual environment. Interest and needs for developing web3D and virtual environment are increasing rapidly. With the help of the RE technique, any real parts can be shared as 3D models in websites or virtual environments. It is also essential in many e-commerce applications.

In summary, reverse engineering is very efficient, has many potential applications, and provides an innovative design paradigm. By combining RE with RP technologies, the cycle for product development can be greatly reduced while the quality is enhanced.

7.6   CASE STUDY

7.6.1   Introduction

The University of Magdeburg, Germany, has derived its name from Otto von Guericke (OVG), the German politician and scientist of the seventeenth century.

There was a need for a number of sculptures in several sizes for awards, souvenirs, and so forth. It was desired that these sculptures be made by casting techniques. Thus, it was determined that the necessary scaled master forms in several sizes should be made by rapid prototyping. Of course, we now know that a digital facet model (STL file) is required for RP.

Therefore, an intelligent reverse engineering approach turned out to be very important for the reproduction of the sculpture of Otto von Guericke for which no design data existed. In the absence of any design data, a computer-aided design file was created using three-dimensional digitizing and surface reconstruction techniques.

7.6.2   Methodology

The general strategies used to reconstruct the sculpture of OVG using reverse engineering, rapid prototyping, and investment casting are illustrated in Figure 7.39.

The geometric data, in this case, was captured using 3D digitizing. The original sculpture was 650 mm high, made from bronze (Fig. 7.40). This was large, heavy, and complex shaped (undercuts, bevels, etc.). The size and weight exceeded the capacity of the local available optical 3D scanners. Therefore, tactile digitizing on a CMM with a probe was performed. The data were captured in 500 radial profiles, creating about 800,000 single data points.

In this situation there were two possibilities for a valid volume representation:

- Surface model or
- Facet model

The shape was too complex for a surface reconstruction. This would require a high number of surface patches. Every patch would be required to be connected to its neighbors precisely. The number, the shape, and the necessary connections would make a surface reconstruction impossible. Also, a CAD internal surface model was not required.

The second possibility was direct polygonization. In a time-consuming polygonization process, a valid STL model was derived from the point cloud.

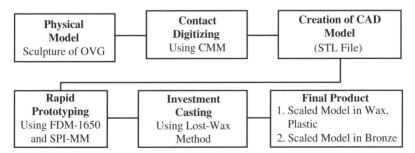

Figure 7.39   General strategies of the proposed approach.

Figure 7.40 Original sculpture.

Many defects inside the point cloud made the polygonization very difficult. Typical errors in the polygonized data were

- Holes
- Gaps between facets
- Deformed facets
- Overlapped facets

The original count, 217,000 facets, was too large to handle with the existing RP machine software. In a stepwise iteration a reduction to 30,000 facet was reached.

The first prototype was made using laminated object modeling (LOM). The shape and quality were examined and some suggestions were made for modifications. With digital sculpturing the changes in styling were made inside the point cloud. Some variants are shown in Figure 7.41.

Using the procedures outlined in the manuals of Sanders Prototype Inc. (SPI) ModelMaker-6B and FDM-1650 RP machine, prototype models of OVG were made in wax and ABS plastic, respectively. Figure 7.42 shows the prototype of OVG in wax.

Figure 7.41 Variants in LOM prototype.

7.6.3 Bronze Cast of OVG

Another objective of the project was to create the sculpture of OVG in metal. An investment casting method was used to make a bronze cast of OVG. Figure 7.43 shows the model created using the following procedure.

- Dissolve the support wax in an ultrasonic cleaner filled with Pertrofirm solvent.

Figure 7.42 Rapid prototype of OVG.

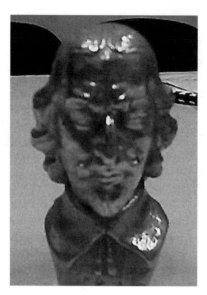

Figure 7.43 Investment cast model in bronze.

- Attach sprues (wax rods) to the wax model.
- Weigh the wax and calculate the amount of metal that would be needed for casting.
- Attach the wax model and sprues to a rubber bottom and secure this to a steel flask cylinder.
- Mix the powder investment and water together.
- Coat the wax with a solution to break the surface tension.
- Pour the liquid investment into the flask and vacuum this solution to remove the air bubbles.
- Allow the investment to harden, thereby creating a ceramic shell in the "green" state.
- Place the flask in a kiln, and over 112 hr, raise the temperature to 1300°F.
- Lower the temperature to 900°F.
- In a hand-held crucible, melt 2 ½ lb of brass with an oxyacetylene torch.
- When the metal is molten, place the flask in a vacuum caster and turn on the vacuum.
- With vacuum running, pour the metal into the flask while keeping a reduction flame on the metal.
- Allow the metal to completely harden.
- Submerge the hot flask in water, breaking apart the investment shell.
- Saw the sprues off, flush to the model.
- Place the metal part in a vibrating polishing machine for 6 hr.

In this case, as many like it, the RE technique was the only viable method available to reproduce this hand-crafted sculpture without the excessive time and difficulty of "resculpting" the figure in some digitized format.

7.7 SUMMARY

In this chapter, the principles and applications of reverse engineering were introduced and discussed. The chapter began with an overview of measuring devices. Both contact and noncontact types were discussed.

The construction of CAD models from point clouds is an essential component of reverse engineering. Both the curved-based and polygon-based approaches for CAD generation were described in detail.

Another issue of RE is data handling and reduction. While many data-handling and reduction methods are available, 2D and 3D grid-band data reduction methods were described in this chapter. The uniform and nonuniform grid and 3D grid methods were also discussed.

The chapter then portrayed present and future applications of RE. Lastly, in order to better comprehend the principles of RE, a case study was presented, showing a step-by-step procedure that was taken for the reproduction of the Otto von Guericke, for which no design data existed.

PROBLEMS

7.1. Which type of measuring device is more suited to measure a freeform surface: contact type or noncontact type? And why?

7.2. What is the view angle constraint in laser scanning?

7.3. When do you obtain a scattered data as shown in Figure 7.8?

7.4. What others can be measured by an industrial computed tomography device than the internal shape?

7.5. Explain the role of registration in preprocessing of point clouds?

7.6. Among the average, Gaussian, and median methods for noise removal, which one works better for suppressing outliers?

7.7. What is subdivision?

7.8. Briefly explain the marching cube algorithm?

7.9. Laser scanning mechanism uses the triangulation method to get shape information from the physical objects. Describe the principle of triangulation in terms of x, y, and z coordinates of measured points.

7.10. In order to scan a point on the part surface using a laser scanner, several constraints must be satisfied. Describe the view angle, field of view, and depth of field constraint, respectively, in the following figure.

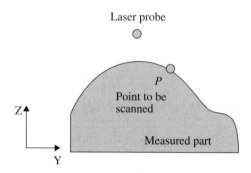

7.11. Determine the coordinate of vertices filtered by AVERAGE method in noise removal if the initial point cloud consist of four vertices such as $P_0(0,0)$, $P_1(1,0)$, $P_2(1,0)$, $P_3(0,1)$ and neighborhood size is 3.

7.12. What is the main difference and characteristics between curve-based modeling and polygon-based modeling?

7.13. The medical images such as MR images and CT images are described by voxels.

   (a) What is the main geometric parameters in the voxel representation?

   (b) What is the material property of the CT image?

   (c) Discuss possible applications of medical images in conjunction with reverse engineering techniques.

7.14. The figure below depicts a scanned line by a laser scanner. Determine which points will be remained after one-directional nonuniform-grid method. Suppose we have maximum angle constraint θ (167°). For this problem you can ignore the maximum grid size condition.

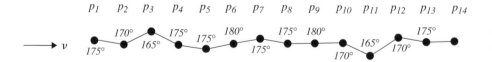

7.15. Suppose we had a spatial bounding box that has 1,500,000 scanned points in it, and we applied the 3D grid data reduction method to the point data. It turned out that the initial grid number was 27 (3,3,3 with respect to x, y, z direction) and the initial grids were subdivided by four times.

(a) How many 3D grids could we have at most when the subdivision process is completed?

(b) What reduction ratio would you expect for this case?

REFERENCES

1. T. Várady, R. R. Martin, and J. Cox, Reverse Engineering of Geometric Models: an Introduction, *Computer-Aided Design,* **29**(4), 255–268, 1997.

2. K. H. Lee, H. Woo, and T. Suk, Point Data Reduction Using 3D Grids, *International Journal of Advance Manufacturing Technology,* **18**(3), 201–210, 2001.

3. N. Ikawa, T. Kishinami, and F. Kimura, *Rapid Product Development,* Chapman & Hall, 1997, pp. 323-328

4. B. L. Curless, New Methods for Surface Reconstruction from Surface Range Images, Ph. D. Dissertation, Stanford University, 1997.

5. K. H. Lee, S.-B. Son, and H.-P. Park, Generation of Inspection Plans for an Integrated Measurement System, The 5th Biennial Conference on Engineering Systems Design & Analysis, Montreux, Switzerland, July 10–13, 2000.

6. *http://www.renishaw.com.*

7. *http://www.brownandsharpe.com.*

8. K. H. Lee, H. P. Park, and S. B. Son, A Framework for Laser Scan Planning of Freeform Surfaces, *International Journal of Advanced Manufacturing Technology,* **17**(3), 171–180, 2001.

9. *http://www.jaewoo.com.*

10. *http://www.aracor.com.*

11. *http://www.siemensmedical.com.*

12. K. Lee, S. K. Lee, and S.-M. Kim, Development of a Universal Fixture for Laser Scanning, *International Journal of Advanced Manufacturing Technology,* Vol. 19, No. 6, 426–431, 2002.

13. N. Amenta, M. Bern, and M. Kamvysselis, A New Voronoi-based Surface Reconstruction Algorithm, SIGGRAPH 98 Proceedings, 1998, pp. 415–422

14. M. Eck, T. De Rose, T. Duchamp, H. Hoppe, M. Lounsbery, and W. Stuetzle, Multiresolution Analysis of Arbitrary Meshes, SIGGRAPH 95 Proceedings, 1995, pp. 173–182.

15. H. Hoppe, T. De Rose, T. Duchamp, J. McDonald, and W. Stuetzle, Surface Reconstruction from Unorganized Points, SIGGRAPH 92 Proceedings, 1992, pp. 71–78.

16. M. Garland, Quadric Based Polygonal Surface Simplification. Ph.D. dissertation, Carnegie Mellon University, 1999.

17. C. T. Loop, Generalized B-spline Surfaces of Arbitrary Topological Type, Ph.D. dissertation, University of Washington, 1992.

18. K. H. Lee, J. H. Ryu, and H. S. Kim, *Contour Based Algorithms for Generating 3D Medical Model,* Numerisation 3D, Paris, 2001 (*http://cadcam.kjist.ac.kr*).

19. W. E. Lorensen, and H. E. Cline, Marching Cubes: A High Resolution 3D Surface Construction Algorithm, *ACM Computer Graphics,* **21**(4), 163–169, 1987.

20. R. R. Martin, I. A. Stroud, and A. D. Marshall, Data Reduction for Reverse Engineering, RECCAD, Deliverable Document 1 COPERNICUS project, No. 1068, Computer and Automation Institute of Hungarian Academy of Science, January 1996.

21. K. H. Lee, H. Woo, and T. Suk, Data Reduction Methods for Reverse Engineering, *International Journal of Advanced Manufacturing Technology,* **17**, 735–743, 2001.

22. K. H. Lee, H. Woo, and T. Suk, Point Data Reduction Using 3D Grids, *International Journal of Advanced Manufacturing Technology,* **18**, 201–210, 2001.

23. W. Schroeder, K. Marin, and B. Lorensen, *The Visualization Toolkit: An Object-Oriented Approach to 3D Graphics.* Prentice Hall, Englewood Cliffs, NJ, 1997.

24. M. Sonka, V. Hlavac, and R. Boyle, *Image Processing, Analysis, and Machine Vision,* PWS Publishing, 1999.

25. *http://www.capture3d.com/.*

26. *http://www.rapidform.co.kr.*

27. *http://www.pixar.com/.*

28. *http://www.geomagic.com.*

8

RAPID TOOLING

Rapid tooling (RT) is a new technique driven by rapid prototyping. The need for faster, better, and less expensive tooling solutions has resulted in the developments of many rapid tooling processes. This chapter will describe both the indirect and direct methods of rapid tooling, followed by a case study on sheet-metal forming by rapid tooling.

8.1 INTRODUCTION

The term *rapid tooling* is typically used to describe a process that either uses a rapid prototyping (RP) model as a pattern to create molds quickly or uses the rapid prototyping process directly to fabricate tools for a limited volume of prototypes [1]. RT also refers to mold cavities that are either indirectly or directly fabricated using the RP technique. There is tremendous interest in rapid tooling solutions these days for product design and manufacturing. Whether RT is used for prototype, short-run, or production tooling, it offers an opportunity to reduce both time and expense of product development.

There are many advantages of rapid tooling for manufacturing. Some of the advantages are as follows:

Shorten the Tooling Lead Time Normal lead time is shortened from months to a few days or weeks.

Low Cost Cost is reduced due to the shortened lead time, so that real trials are more affordable.

Functional Test of Parts in Early Design Made Possible   Due to short tooling lead time and low cost in using RT, many engineers prefer to produce parts for functional tests. Most of the faults are debugged before production, thereby avoiding many design disasters.

Direct Transfer of Parts in Early Design Made Possible   Many careless human mistakes due to misinterpretation of drawings can be avoided because the original CAD drawing can reveal errors in the early stages of the RT process.

Rapid tooling can be classified into indirect and direct methods of rapid tooling as shown in Figure 8.1. The indirect method uses RP master patterns to produce a mold, while the direct method involves building the actual core and cavity mold inserts by an RP machine. Silicon rubber molding and epoxy tooling are some examples of indirect methods. In some direct methods, the RP process can produce metallic and ceramic tools such as in selective laser sintering (SLS) and multiphase jet solidification (MJS) [2]. In the next section, these two types of RT will be presented.

8.2   INDIRECT METHODS OF RT

The demand for faster, more accurate tooling solutions has resulted in the development of many RT methods. Some of these methods have been commercialized while others are still in the conceptual or developmental stages. Each RT method has its own merits and limitations. However, many companies were willing to experiment with the RT process because of tremendous market potential.

This section presents several pattern-based processes that have been developed for creating molds rapidly, with varying costs, lead times, and process capabilities. The accuracy of these processes depends in part on the accuracy of the RP process used to create the pattern. Typical indirect systems along with some pertinent information are shown in Table 8.1 [3].

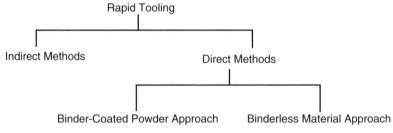

Figure 8.1   Classification of rapid tooling processes.

TABLE 8.1 Commercially Available Indirect Tooling Processes

Process	RTV Silicone Rubber Mold	Aluminum-Filled Epoxy	Sprayed Material	Kirksite	3D Keltool
Suppliers	Many	Many	Many	Many	3D Systems & licenses
Lead Time	0.5–2 weeks	1–4 weeks	2–4 weeks	3–6 weeks	1–6 weeks
Applicable Quantities	10–50	100s complex 1000s simple	50–1000s	50–1000s	50 to millions
Relative Cost	$1K–$5K	Typ $3000; range: $2.5–$10K up to $35K	$2–$15K	$4–$15K	$2 to $5K
Materials	Urethanes, epoxies, acrylics	Thermoplastics	Thermoplastics	Thermoplastics	Thermoplastics
Tolerance (in./in.) or as Designated	±0.005 w/0.020 walls	± 0.002 in./in.	± 0.002 in./in.	± 0.003 in./in.	± 0.001; flatness ± 0.025 mm/in.
Hardness	N/A	N/A			Rc 32 heat trt to Rc 42–46
Mold Parameters					Max temp 650°C; 20K–25K
Surface Finish					20–25 μin.
Part Size Limitations					5 × 8.5 × 5 in.
Strengths	Least expensive mold (but fairly expensive piece parts)	Least expensive for true thermoplastics	Large parts	Complex shapes	Accuracy, high volume
Weakness	Tool life; accuracy better for simple parts; limited materials	Long cycle times, tool life; accuracy, better for simple parts (CIC)	Tool life; accuracy, better for simple parts; poor for narrow slots	Accuracy	Part size, one supplier

8.2.1   Room Temperature Vulcanizing Silicone Rubber Molds

The production of room temperature vulcanizing (RTV) silicone rubber molds has become one of the most popular types of RT applications [4]. Silicone is used because of its ability to be molded around a master pattern to produce a cavity. With the arrival of rapid prototyping techniques, master patterns are often made of the RP models themselves. A detailed silicone RTV rubber molding procedure is shown in Figure 8.2, even though the steps taken are rather simple [5].

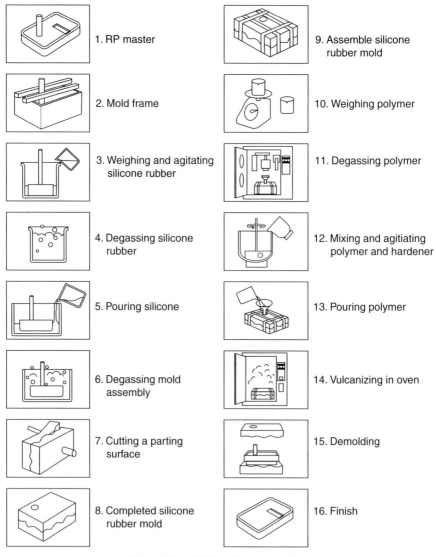

Figure 8.2   RTV rubber procedure.

Two-part thermoset materials such as polyurethane can be molded within the cavity. These materials are available in a variety of mechanical properties and can mimic the mechanical and thermal properties of elastomers, ABS, nylon, and other popular thermoplastics. Polyurethane is usually poured into the silicone rubber cavity under vacuum to avoid any formation of air bubbles within the molded component. The silicone rubber tool will generally produce about 20 polyurethane parts before it begins to deteriorate. However, this depends on the amount of detail in the tool and the type of polyurethane being molded [4].

Silicone rubber tooling provides fast, inexpensive molds, excellent part cosmetics, and the option of using multiple materials. The process is suitable for small or large parts. The primary weakness of the process is that the properties of the urethane materials are different from those of the thermoplastic materials used in production. Due to material cost and labor demands, individual part prices are relatively high. Even with its limitations, silicone rubber tooling can be used as a production process.

8.2.2 Spray Metal Tooling

Metal spraying is used for the production of soft tooling [4]. Model preparation is the first and one of the most important steps in this process. Depending on the finish of the model, sanding should be done because all surface imperfections become apparent in the sprayed shell. Typically, the master (SLA, LOM, etc.) must be hand finished to the desired quality before the mold is made. In most cases, a silicone mold and urethane reproduction are made for the tooling master because chances are this master will be destroyed. The process involves spraying a thin shell of about 0.080-in. (2-mm) thickness over a pattern and backing this with epoxy resin to give it rigidity. The metal spraying is done using a variety of techniques. With most RP techniques, the models produced have a low glass transition temperature (i.e., the temperature where the material starts to change to a soft amorphous structure). Therefore, it is important to keep the pattern temperature as low as possible when spraying. If the temperature of the model gets too high, it will start to relax and distort, which then results in an inaccurate tool. An alternative is to use the RP pattern to create a silicone rubber mold, which is then used to create a ceramic spray substrate. Although the ceramic substrate can withstand the high-temperature metal spray, this process increases time and cost. An example of the spray metal molding process using an RP&M model is shown in Figure 8.3 [5].

The most popular techniques for use with RP models are spraying low-melting-point alloys (lead/tin based) with a gun similar to a paint sprayer and metal deposition with an arc system. In the wire-arc spray, the metal comes in filament form [6]. The arc system feeds two wires into a gun and an electric arc is struck between them. This causes the wire material to melt, and then a compressed gas atomizes and sprays it onto the pattern. The higher the melting point of the wire material, the more difficult it is to keep the pattern

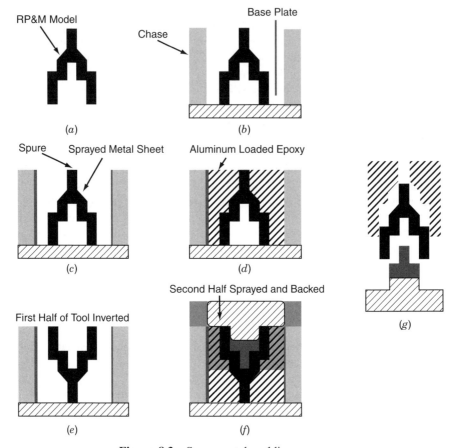

Figure 8.3   Spray metal molding process.

cool. Therefore, it is common to spray zinc- or aluminum-based alloys directly onto RP models. One technique is to apply a metallic coating by using electrolysis plating or physical vapor deposition. Once there is a metallic coating on the model, heat will be transmitted more readily across its surface.

A problem associated with metal spraying is that it produces shells with high internal stresses. It is possible to counteract these by simultaneously shot-peening the sprayed shell. Steel shot fired at the shell during spraying induces compressive stresses that counteract the tensile stresses. Also an aluminum or steel frame can be fabricated to absorb the pressures of molding and to allow the completed mold to be installed in the molding equipment. These materials will absorb most of the compression produced by the machine.

Metal spraying is typically used on models that have large gently curved surfaces because it is difficult to spray into narrow slots or small-diameter holes. If a model has such features, then it is common to create brass inserts and locate the features. Then the model is sprayed around the inserts. When the model is removed from the shell, the inserts are permanently fixed into the shell. These inserts are stronger than the shell material, which is weak and breaks easily if formed as a tall, thin feature.

Spray metal tools can produce more than 1000 parts depending on the process, material being formed, and the amount of care given to the tool. Clamping and injection pressures for metal-sprayed injection tools are usually less than those for steel or aluminum tools and may affect the mechanical properties of the injection-molded part. With the shell being very thin and generally backed up with an epoxy-based resin, the thermal conductivity of a metal-sprayed tool is less than that of an aluminum or steel tool. This also affects the mechanical properties of the injection-molded components and can increase cycle time. Some plastics are much more corrosive and abrasive on tool faces but can be partially overcome by a variety of techniques, such as plating the tool surface with nickel or chrome, or using aluminum or steel inserts.

Spray metal tools have been used in many applications including sheet-metal forming, injection molding, compression molding, blow molding, and prepreg sheet layup. Various plastics have been molded including polypropylene, ABS, polystyrene, and difficult process materials such as reinforced nylon and polycarbonate [4]. The main advantage of spray metal tooling is the ability to produce large tools quickly. However, it may be difficult or impossible to spray into narrow slots or deep holes, meaning that the part geometry must be relatively simple. Molds are not particularly strong and the process requires special equipment and a special operating environment.

8.2.3 Sprayed Steel

The Sprayform sprayed steel process is similar in method to traditional sprayed metal tooling in that atomized material is deposited using a spray gun [4]. However, the main difference lies in the mechanization, which uses multiple spray heads throughout the process. The process produces much harder tools and is, therefore, a much more useful process than traditional sprayed metal tooling. Molds created with the process have been found to last as much as 20% longer than comparable machined molds of the same materials. After the pattern has been coated with sufficient material, it is removed and the resultant metal block undergoes a final machining to fit in a mold frame.

The primary advantage of the Sprayform process is that it works well for large tools, especially as sheet-metal stamping dies. It offers a high deposition rate and is less expensive compared to conventionally machined steel tooling. The cost of the equipment and the licensing fees, as well as the limitation of spraying into holes and slots, are limiting factors. Currently Ford Motor Company is developing equipment that allows it to rapidly manufacture production tools and dies, instead of only using the technology for prototypes.

8.2.4   Cast Aluminum and Zinc Kirksite Tooling

Cast kirksite cavities from SLA models can provide excellent rapid tools for prototype and bridge to production injection-molded parts. Parts can be molded in 2 to 3 weeks in any thermoplastic production resin [7]. Using molten metal casting techniques, it is possible to cast around an accurate pattern with either aluminum- or zinc-based alloys [4]. This type of technique allows for higher production volumes and the use of more aggressive polymers that give the tool a high degree of mechanical hardness.

Given the casting temperature of both aluminum and zinc, it is important to replicate the initial pattern into a material capable of withstanding such heat. Using silicone tooling, as detailed earlier, a cavity is produced around the model. However, rather than a resin copy being made, the silicone cavity is then filled with ceramic. After drying, the ceramic prototype is placed into a bolster and covered with the molten metal.

Cast aluminum and zinc kirksite tooling offer a simple and low-cost method of tooling. However, because the master file must include the appropriate shrink factors for both the kirksite and plastic material, some loss of tolerance is to be expected. The process is capable of producing parts perfectly acceptable for functional use, but not at the level of tolerance as those produced from a machined steel mold yet [7].

8.2.5   Three-Dimensional Keltool

The 3D Keltool process typically starts with a CAD design of the core and cavity mold inserts, followed by the creation of the core and cavity patterns with stereolithography or some other RP process [4]. Once these core and cavity patterns have been finished to the desired surface, silicone rubber is cast against them to create molds into which a mixture of metal powder and binder is poured, packed, and cured. The metal mixture consists of finely powdered A6 tool steel and even finer particles of tungsten carbide. At this point, the cast core and cavity inserts exist in a *green* state. These green inserts are fired in a hydrogen reduction furnace to burn away the binder, sinter the metal particles, and infiltrate copper into the inserts. This produces solid metal inserts that are approximately 70% steel and 30% copper with physical properties similar to that of P20 tool steel. The inserts are finish machined, drilled for ejector pins, and fitted into mold bases. The tools from this process show very good definition and surface finish.

Although the 3D Keltool process generally starts with a design of the finished part, the process can be reversed to provide an stereolithography pattern of the core and cavity instead. Both pathways are represented in Figure 8.4 shown below [5].

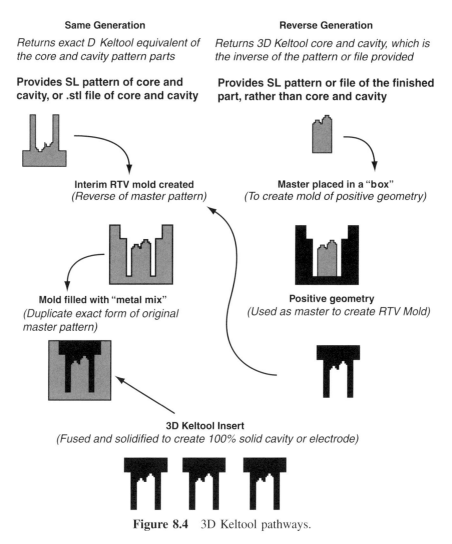

Same Generation

Returns exact D Keltool equivalent of the core and cavity pattern parts

Provides SL pattern of core and cavity, or .stl file of core and cavity

Reverse Generation

Returns 3D Keltool core and cavity, which is the inverse of the pattern or file provided

Provides SL pattern or file of the finished part, rather than core and cavity

Interim RTV mold created
(Reverse of master pattern)

Master placed in a "box"
(To create mold of positive geometry)

Mold filled with "metal mix"
(Duplicate exact form of original master pattern)

Positive geometry
(Used as master to create RTV Mold)

3D Keltool Insert
(Fused and solidified to create 100% solid cavity or electrode)

Figure 8.4   3D Keltool pathways.

8.2.6   Vacuum Casting

Vacuum casting is a copying technique used for making small series of functional plastic parts. A master model (that typically originates from stereolithography or selective laser sintering) is carefully prepared to ensure a high-quality finish to the surface and the definition of the parting planes. Silicone is cast around the master, partially under vacuum in order to avoid air bubbles being trapped in between master and silicone [8]. The two-part resin is mixed and degassed before being poured into the silicone cavity. After pouring, the vacuum is released and the tool is removed to a postcuring oven for up to 2 hr depending on tool size. Following an exothermic reaction of the two-part resin, the cavity is opened and a polyurethane part removed. The silicone cavity is then closed and the process repeated.

8.2.7 Reaction Injection Molding

Reaction injection molding (RIM) is a process that uses a simple resin injection system with two pressurized chambers [4]. The process takes its name from a chemical reaction that occurs within the tool. The plastics used are thermosets, either polyurethanes or foamed polyurethanes [7]. The two components that produce the polyurethane are mixed just prior to injection into the tool. The silicone tool is filled with an injection nozzle at atmospheric pressure until excess resin is driven up through a series of riser holes. The cure reaction time is much shorter than in vacuum casting. With the low viscosity and low injection pressures, large, complex parts can be produced economically in low quantities. Considerable design freedom is possible, including thick- and thin-wall sections that are not good for injection molding due to the uniform shrink characteristics. Foamed polyurethanes are natural thermal and acoustic insulators [7]. Excellent flowability allows for the encapsulation of a variety of inserts. There is no thermal cycling and the contact time between the resin and silicone rubber is much shorter; therefore, the tools can last for up to 100 shots.

8.2.8 Wax Injection Molding

The silicone cavities used in two-part resin molding are also fitting for low-pressure injection molding of low-molecular-weight paraffin waxes for investment casting [4]. The silicone cavity is filled with semimolten wax using a low-pressure injection system. The cavity is then chilled until the wax has fully solidified, at which time the wax is removed, and the process is repeated. The investment casting wax material is very fragile, and extreme care must be exercised when removing wax parts from the silicone tool.

8.2.9 Spin Casting

Spin casting offers a fast and relatively inexpensive method for casting high-strength engineered metal parts. Parts or models are laid out on a disk of uncured silicone rubber. Depending upon model/pattern thickness and shape, cavities may be cut or molded by hand to accommodate the part. The uncured silicone material is soft and pliable like clay [9]. The mold parting line is formed at this stage and can be built up or lowered around any section of the model/pattern. Cores and pull-out sections can also be easily incorporated, if required. Because the tools used in the spin-casting process are made from vulcanized rubber, the process makes it possible to cast a range of materials from polyurethanes to zinc-based alloys [4]. The cavities are filled using centrifugal force to pressurize the tools cavity. This process is ideal for forming small zinc castings that will ultimately be produced by die casting. Spin-cast tools can produce in excess of 100 replicated parts before degradation of the tool. Parts also are produced in a range of low-melting-point alloys.

The advantages of spin casting include the option of processing a variety of materials that include: zinc, lead, and tin alloys to polyurethanes, polyesters, epoxies, and elastomers [9]. The process is relatively quick and the equipment is relatively inexpensive. Some disadvantages include size limitations and strength of materials.

8.2.10 Cast Resin Tooling

Cast resin tooling is a straightforward and economical method of producing a tool for the injection molding of thermoplastic parts [4]. The process has been around since the 1940s; however, with the advent of RP there has been a reintroduction of cast resin tooling. It is generally a four-step procedure consisting of:

1. Mounting the pattern within a mold box
2. Setting up a parting line
3. Painting
4. Pouring resin over the pattern (until there is sufficient material to form one half of the tool)

After completing the first half, the process is repeated for the other half of the tool. There are many tooling resins available with different mechanical and thermal characteristics. Epoxy is considered to be the most commonly used. The resins are often loaded with aluminum powder or pellets to improve the thermal conductivity and compression strength of the tool and to reduce the cost of the resin. To achieve the desired properties, the tools may need to be postcured at an elevated temperature. For high-pressure molding processes, such as injection molding, cast resin tool inserts are usually placed within steel bolsters to restrain the clamping/injection force [10].

Cast resin tools are usually used for 100 to 200 molded parts, although it is possible to get up to 1000 parts, depending on the material being molded. This process is fast, relatively simple, and can be used to mold common thermoplastics such as polypropylene and ABS. However, low mechanical strength of the molds makes this method of rapid tooling suitable for only relatively simple shapes. Care must be taken during curing, especially with larger parts, to avoid excessive distortion due to exotherm of the resin. Lead times of about 10 days are common with production volumes of 100 parts.

8.2.11 Rapid Solidification Process

Rapid solidification process (RSP) differs from other sprayed metal processes in that it can deposit hundreds of pounds of material per hour, while the conventional wire-feed systems deposit approximately 15.4 lb (7 kg) per hour

[4]. The rapid solidification process bypasses the majority of conventional fabrication steps, which helps to save both time and money. A high-velocity jet of inert gas sprays tiny droplets of molten metal onto a pattern to be replicated. After depositing the spray, the pattern is removed, the deposit is trimmed to fit a standard mold base, and is heat treated, if necessary. The surface area of the droplets is so great compared to their volume that the droplets cool somewhere between 100° and 100,000° per second [11]. Because the droplets cool at an extremely fast rate, this results in unusual beneficial characteristics of the alloy. The result is a mold that resists distortion during heat treatment and last about 20% longer in production runs. The RSP process could potentially be used to build the entire tool as opposed to a thin shell that requires back filling. A current drawback is the size limitation of about only 6 in. (150 mm).

8.2.12   Plaster Molds

Plaster mold casting, also called rubber plaster molding (RPM), is a method of producing aluminum or zinc castings by pouring liquid metal into plaster molds and offers an alternative to investment casting, sand casting, and prototype die casting [7]. Although there are several variations of this process, it usually begins with the creation of the model or master pattern. A silicone rubber reversal is then molded over the master. A second silicone rubber is molded into the first in order to provide a silicone rubber positive of the original model. The plaster cavity is created by molding plaster around the second silicone rubber positive. The molten metal is then poured into the plaster cavity. Once the metal has solidified, the plaster is broken away and the tool is complete. Any required machining or heat treatment can be performed once the plaster has been broken away from the tool.

The rubber version of the master is required so that it can easily be withdrawn from the plaster mold. It is also possible to mold epoxy off of the master and pour plaster over this [4]. The epoxy molds will have a greater life than those made from rubber.

The plaster mold process has a low mold cost and good surface detail. Also, it is possible to produce reasonably large parts with this process. However, because the material has lower cooling rates, the mechanical properties are somewhat lower than normal. This can lead to parts with a yield strength that is 20% lower than conventional die casting. It should be realized that the materials used for plaster casting slightly differ compared to die casting.

8.2.13   Electroforming

Electroforming involves electroplating a thick shell (several millimeters thick) onto a master pattern [4]. The ability to start plating requires that the surface of the pattern be electrically conductive. Therefore, for surfaces that are not electrically conductive, a conductive lacquer can be sprayed onto the model.

Once the plating has been done, the shell is removed and then backed with a suitable material.

The technique of electroforming can be used to produce tools for shoe soles with complicated patterns from original wax models. Nickel is a common material for electroforming because it has good thermal conductivity and strength. The process gives faithful reproductions of the master but can be limited when plating into deep narrow slots or holes. Sometimes electroplating can build up more material on exterior corners, causing narrow slots to become closed at the top before they have been properly plated at the bottom. This problem can be partially overcome by reducing the current, however, this increases the amount of time to produce the shell.

Express Tool (Warwick, Rhode Island) is developing an electroforming process that it plans to commercialize. The company typically produces patterns by machining graphite. This material is an excellent conductor and it machines many times faster than aluminum. Another benefit to using graphite is that it serves as a natural release due to its lubricant properties—making it easier to separate it from the nickel shell. The process of electroforming is not particularly fast, but for objects without deep slots and holes, a good reproduction of detail can be achieved.

8.2.14 Investment Cast Tooling

Investment casting has been used with RP models to produce metal tooling [4]. Traditional methods of investment casting with injection molding is compared with investment casting using RP. As shown in Figure 8.5, RP saves time by eliminating the need to manufacture the injection mold [5].

Most of the tools are cast from aluminum, but some steel tool molds have also been used. If a steel or hardened alloy cavity is required, either for mechanical strength and thermal cycling or due to high-volume production,

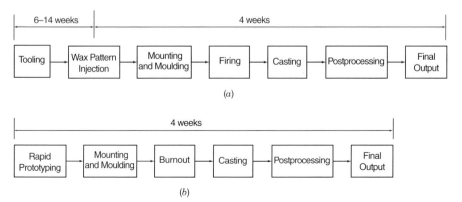

Figure 8.5 Investment casting process: (*a*) traditional method and (*b*) RP&M part.

then the kirksite process becomes an option. The kirksite process offers an alternative to open-cast tooling. The lost-wax process can be used to replicate the part in metal by first making a sacrificial RP model of the desired cavity.

The RP pattern is first invested in multiple layers of ceramic slurry, which are allowed to dry between coats. After the shell has dried, the ceramic shell and invested part are fired. The firing process sinters the ceramic shell and causes the invested model to be burned out. After firing, any ash residue is washed away from the ceramic shell. The molten alloy of the tool material is then poured through a gating system into the void left by the RP pattern. After solidification and cooling, the ceramic shell is fractured and the newly formed metal cavity is removed and postprocess machined.

Investment cast tools have been used for injection mold cavities and die casting tools. However, due to the unpredictable contraction of the casting process, it is difficult to maintain a high level of accuracy with this tooling process.

8.3   DIRECT METHODS OF RT

The methods of rapid tooling described above involved the indirect production of a master pattern from which the tool is produced. One of the major concerns of producing these tools is the amount of time it takes to complete the pattern. In addition, past replication techniques have been known to carry inaccuracies. Companies would ultimately want to produce tooling directly, although there are currently limitations to most direct tooling methods [4].

This is where the additive "layer manufacturing" techniques come in. These techniques allow for unconventional features to be integrated into the tool that would otherwise be impossible to achieve with conventional tooling techniques. The most significant feature is conformal cooling (or heating) channels that allows for the cooling or heating of the tool at points where it is needed, instead of only where the channels can be conveniently drilled. Study of this matter has shown that conformal channels can reduce injection mold cycle times by 40%. Table 8.2 shows the commercially available direct tooling processes [3].

8.3.1   Direct ACES Injection Molding (AIM) Tooling

In the AIM process, the mold is "grown" using the SLA process. 3D Systems of Valencia, California, came up with the process known as direct AIM [ACES (accurate clear epoxy solid) injection molding] [4]. With the help of direct AIM, injecting a range of thermoplastics into cavities and producing usable parts is possible. The mold is similar to a regular part SLA but is the negative image and cut into two halves. The process works by taking a part geometry, designing injection molding inserts around it, and creating the inserts in the SL system. Then the molten plastic is shot into the inserts to

TABLE 8.2 Commercially Available Direct Tooling Processes

Process	Direct AIM	Copper Polyamide SLS	Direct Metal Laser Sintering	RapidTool	DirectTool
Suppliers	3D Systems, SBs	3D Systems, SBs	EOS BmbH, SBs	3D Systems, SBs	EOS GmbH, SBs
Lead Time	1 week	1–5 days	1–4 weeks	2–5 weeks	1–2 weeks
Applicable Quantities	10–50	1–500	100–1000	100s die cast, 100,000 plastics	100s die cast, 100,000 plastics
Relative Cost (thousand)	$2–$5			$4–$10	
Materials	Low temp., unfilled thermoplastics	Thermoplastics	Thermoplastics	Thermoplastics, metals	Thermoplastics, metals
Tolerance (in./in.)	± 0.002		± 0.002	0.003 layers, 0.005 details,	± 0.001–0.002
Hardness		Shore D-2240 < 500°F		Rb 87	Brinell 60 - 80
Mold Parameters	Requires experimentation and experience			Injection molding pressure and temperature	
Surface Finish		500 μin., as processed; 800 μin. after finishing		5 μ; 1–3 μin. after polishing	$R_z = 20$ μm, shot peened
Part Size Limitations		10 × 10 × 6 in.		8 × 10 × 5 in.	10 × 10 × 7 in.
Strengths	Direct fabrication of mold	Close to hard tool cycle time and temp.; conformal cooling; no burnout cycle	Conformal cooling; no burnout cycle	Die cast; withstands high mold temperature and pressures	No burnout cycle; accurate; surface finishes are improving
Weaknesses	Severe material and process limitations	Limited tool life; lower pressures; conformal cooling channels have limitations	Limited tool life; lower pressures	Requires burnout and infiltration cycle; may require finishing	May require finish machining

create the part. Currently, only less abrasive and lower melting point polymers are being used, but researchers promise improved materials for these applications in time. In some cases, molded parts can be produced within 4 to 5 days of receiving a CAD database, which makes this process fast.

Stereolithography tools generally are produced of the standard commercially available stereolithography resin. However, the cavity can be filled with a variety of materials including thermoplastics, aluminum-filled epoxy, ceramics, and low-melt temperature metals. As many as 500 parts can be molded from a single tool, although the typical production is from 10 to 50 parts. Due to rapid production and the thermoplastics being used, low tool strength and risk of failures are current disadvantages to this process.

8.3.2   Copper Polyamide Tooling

The copper polyamide tooling process from DTM (Austin, Texas) involves the selective laser sintering of a copper and polyamide powder matrix to form a tool [4]. Sintering occurs between the polyamide powder particles to create actual molds complete with cavities and core. No furnace processing is necessary, and the composite mold inserts are machinable. For a relatively simple geometry with no slides or side cores, you can get tools in less than a week. The process boasts an increase in tool toughness and heat transfer over some of the other soft tooling methods. The copper provides these beneficial characteristics along with allowing for the running of a tool with pressure and temperature settings that are closer to production settings. The lower material strength continues to be the primary disadvantage in copper polyamide tooling.

8.3.3   Direct Metal Laser Sintering

Direct metal laser sintering (DMLS) from EOS uses direct processing of metal powders with laser sintering machines [4]. The machine is used for the production of tool inserts, but it also is possible to produce metal components. This process uses liquid-phase sintering to bind hard metal particles together using a lower melting point metal as a binder. By optimizing the building parameters, the typical shrinkage due to the sintering is compensated by an expansion of the material caused by diffusion of the components so that this material has no net volume change during the laser sintering process, thereby enabling a very high accuracy [12]. Most importantly, many case studies showed that DMLS tools and inserts could be used for injection molding under standard parameters with just some hand finishing of the critical surfaces and did not require postmachining. There are two materials that are available for the DMLS process:

   Bronze-Based Materials   These are used for injection molding of up to 1000 parts in a variety of materials.

Steel-Based Materials These are useful for up to 100,000 plastic injection-molded parts.

The first stage of the process chain is essentially the same as for the conventional process (NC milling), although to gain the most benefit from this process chain, certain factors should be considered when designing the tool. For example, laser sintering large volumes of regular geometry that could be more efficiently milled should be avoided, and designing in pilot holes for ejector pins is preferable. The process chain also allows various new possibilities such as designing in conformal cooling channels that can also improve performance. It is not necessary to add machining allowances. The second stage is laser sintering of the mold inserts, which typically takes 1 to 2 days, depending on the size. For special applications such as blow molding or vulcanization, the DMLS parts can be used without any postprocessing, but for injection molding applications it is usually necessary to infiltrate the part to fill the surface pores. The preferred infiltration fluid is a high-temperature epoxy resin because this has no influence on the accuracy due to the relatively low heat impact to the metal. DMLS offers good feature definition, although the surface definition of the steel-based powder builds more slowly. The current finest version of bronze-based powder and steel-based powder builds with 50-μm layer thickness [12].

8.3.4 Selective Laser Sintering RapidSteel

Similar to the way cavities are generated directly through stereolithography, it is also possible to build tool cavities using the laser sintering process [4]. DTM has come up with RapidSteel (aka RapidTool), where digital models of the core and cavity geometries are created and sent to a Sinterstation machine for fabrication in RapidSteel powder. The metal powder consists of 316L stainless steel and bronze and is mixed with a polymer binder system. The laser solidifies the binder one slice at a time forming the part. The metal powder is suspended in the binder. This step is commonly referred to as the green part. The green parts are then fired in a furnace in order to remove the polymer binder and infiltrate bronze into the mold inserts through a capillary phenomenon. The product becomes a dense tool that consists of 40% bronze and 60% steel. Finally the inserts are finished; ejector pins are drilled and fit to a mold base. Build times vary depending upon the particular model Sinterstation you are operating.

The product is a durable mold with good tool strength that can be used for injection mold tooling, as well as die casting applications. RapidSteel molds have been used for hundreds of cast aluminum, zinc, and magnesium parts. The RapidSteel 2.0 material is an excellent solution for generating mold inserts capable of producing 100,000 injection-molded parts—or several hundred Al, Zn, or Mg die cast parts [13]. RapidSteel mold inserts have strength and hardness properties superior to aluminum tools and are highly thermally conductive. These tools have wear characteristics similar to traditional steel tools—and can be plated, textured, welded, and polished. RapidSteel allows users to create complex geometry and molds that can withstand conditions of injection molding. However, RapidSteel requires machining and polishing finish time, in addition to requiring cost of equipment.

8.3.5 Laminated Tooling

Laminated tooling is an alternative method to building cavities directly with an RP machine that uses principles similar to laminated object manufacturing (LOM) [4]. In this process, layers of sheet metal are cut to replicate slices through a CAD model, using laser cutting or water jet technologies.

In order to create a mold tool, the CAD model must take the form of the required cavity. By cutting all of the slices of the cavity in sheet metal, a stack of laminates can be made to replicate the original CAD model. Clamping or diffusion bonding makes it possible to create a pseudosolid cavity in hardened tool steel without the need for complex postprocess cutter path planning. Due to the use of relatively thick laminates—typically 0.040 in. (1 mm)—the surface finish is poor, requiring a machined finish. The tool life is a function of the initial sheet material, which can be hardened after cutting and lamination. However, part complexity is bounded by layer thickness. One significant advantage of laminated tooling is the ability to change the design of parts quickly. Cooling channels could be easily incorporated within the tool design, and laminated tooling is good for large tools as well. Laminated tools have been used successfully for a variety of techniques including press tools, blow molding, injection molding, and thermal forming.

Research also is being performed in the use of laminate tools in pressure die casting. The sheet material can be hardened after cutting and lamination in order to enhance tool life. However, part complexity is bounded by layer thickness.

8.3.6 Laser Engineered Net Shaping

The laser engineered net shaping (LENS) system from Optomec (Albuquerque, New Mexico) was originally developed at Sandia National Laboratories [4]. Parts were built using a metal powder feed and laser cladding. The high-power Nd : YAG laser is used to melt metal powder supplied coaxially to the focus of the laser beam through a deposition head. The laser beam typically travels through the center of the head and is focused to a small spot by one or more lenses. Fabrication process occurs in a low-pressure argon chamber for oxygen-free operation [3]. A motion system moves a platform horizontally and laterally as the laser beam traces the cross section of the part being produced. After a layer has been formed, the machine's powder delivery nozzle moves upward prior to building the next layer.

The LENS process is an additive fabrication method, although it produces fully dense metal parts. Generally, materials such as 316 and 304 stainless steel, nickel-based superalloys such as Inconel 625, 690, and 718, H13 tool steel, tungsten, Ti-6Al-4V titanium alloy, and nickel aluminides have been used in industry. The strength of the process lies in the ability to fabricate fully dense metal parts with good metallurgical properties at reasonable speeds [3]. Objects fabricated are near net shape but generally will require finish machining. They have good grain structure and have properties similar to, or even better than, the intrinsic materials. Selective laser sintering is at present the only other main commercialized RP process that can produce metal parts directly.

8.3.7 Controlled Metal Buildup

Albrecht Röders GmbH & Co KG (Soltau, Germany) commercialized a process known as controlled metal buildup (CMB) [4]. This process was originally developed at the Fraunhofer Institute for Production Technology (IPT) in Aachen, Germany.

The CMB systems use laser cladding and milling that result in fully dense parts. Metal buildup is controlled by laying down a layer of metal with its laser and then milling it flat, and laying down another layer of metal and repeating the milling/lasing cycle until the part is formed or repaired [14]. A steel wire and a 1- to 2-kW HDL laser weld the steel onto the surface of the workpiece. A high-speed cutter then flattens each layer before a new layer is deposited. Any weldable metal can be handled by the 1-kW laser, which is mounted in front of the milling spindle and retracts during the milling cycle.

8.3.8 Prometal

ExtrudeHone's (Irwin, Pennsylvania) ProMetal rapid tooling system (RTS) is the commercial realization of MIT's three-dimensional printing (3DP) process for manufacturing metal parts and tooling [4]. The RTS 300 model is capable of creating 12 × 12 × 10 in. (300 × 300 × 250 mm) steel parts. The RTS system includes the following applications: plastic injection molding, vacuum forming, blow molding, lost foam patterns, and direct fabrication of powder metal components

In 1999, ExtrudeHone sold its first commercial RTS-300 to Motorola. Motorola then joined a collaborative effort consisting of several industrial members, all part of MIT's Three-Dimensional Printing Consortium. The finished structural skeleton is then sintered and infiltrated with bronze to produce a finished part that is 60% steel and 40% bronze [15]. Early reliability problems arose and delayed the implementation effort, but recent advances have produced promising results.

8.4   CASE STUDY: SHEET-METAL FORMING BY RT

A study of sheet-metal forming on tools made by rapid prototyping technologies was conducted in a research program that involved the stereolithography process from 3D Systems, fused deposition modeling (FDM) from Stratasys, and laminated object manufacturing (LOM) from Helisys Corp. [16]. Some of the Helisys composite louver dies made using LOM are shown in Figure 8.6. After the tools were made, the sheet was formed by a simple hydraulic press and by the Guerin process, also known as rubber pad forming.

Forming operations begin with a relatively simple shape—a sheet-metal blank, that is then plastically deformed through one or more operations into a relatively complex configuration. This process drastically reduces metal removal requirements and brings the product to near net or net shape dimensions. However, the process usually requires relatively expensive tooling. Therefore, reducing this tooling cost was the main goals of the study [16].

The cost savings in labor hours through stereolithography were 98% over conventional methods. Therefore, exploring whether such cost savings were possible with other rapid prototyping systems was also considered worthwhile.

8.4.1   Forming Technologies

Sheet metal is formed in numerous variations of the simple hydraulic press. Systems such as the rubber pad forming method covers several types. In this

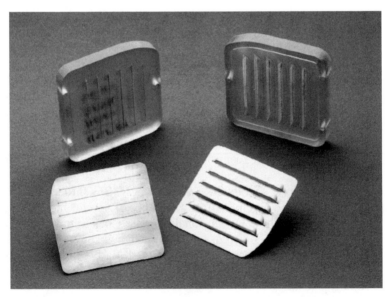

Figure 8.6   Custom louver dies with blank and formed blank made by LOM.

study, success was achieved with the Guerin process of rubber pad forming, as well as with a simple hydraulic press. The Guerin process is the oldest as well as the most basic of the rubber pad processes. In rubber pad forming, the rubber presses the sheet-metal blank around the form block. Other types of rubber pad forming are the Verson-Wheelon and the hydroforming processes, which are slightly more efficient because a flexible hydraulic cell helps adjust the rubber around the workpiece for better forming. These processes have the advantage of not requiring a top die since the flexible rubber adapts to the shape of the bottom die and even allows sheet metal of different thicknesses.

8.4.2   Stereolithography Process

Stereolithography is a process by which a solid model from a CAD is sliced by software into thin layers of approximately 0.1 mm (0.004 in.), with variations from 0.1 to 0.5 mm (0.004 to 0.020 in.) [16]. An ultraviolet laser then scans these layers onto the surface of a light-sensitive liquid plastic (a photopolymer resin). As each layer is cured, it descends by 0.1 mm (0.004 in.) into the liquid, making room for the next layer. Once this process has completed all the layers of the model, it rises out of the liquid as a complete part, accurate to a few thousandths of an inch. QuickCast is a hollow honeycomb build style with a tough outer skin, made of the same photopolymer as the solid parts. It is similar to the lost-wax pattern for investment shell casting.

Stereolithography was used in three different ways to form sheet-metal blanks:

- A stereolithography ACES part was built as a form block (Fig. 8.7).
- A QuickCast part whose internal hatching was filled with a steel-filled epoxy, or a filled epoxy liquid molding compound (LMC), became a curing mold for an epoxy material (Figs. 8.7 and 8.8)
- A stereolithography part became a casting mold, which was then filled (cast) with a steel-filled LMC (Fig. 8.9)

The resulting parts were tested at 350 tons on a rubber pad forming press with 1-mm (0.040-in.) thick 7075-0 condition annealed aluminum. All the form blocks held up without damage.

The ACES form block, made of Ciba Specialty Chemicals SL5170 resin, can create 30 to 50 sheet-metal parts. The compressive strength of the SL5170 epoxy resin is 82 MPa (11,700 psi). Ciba Geigy also provided the compressive modulus: 1965 MPa (285,500 psi), the hardness resistance to bend. This test was done on an ACES block $12.7 \times 6.8 \times 25.4$ mm ($0.5 \times 0.27 \times 1$ in.), simulating the ASTM 695 standard.

The most notable work was the creation of four different custom louver-forming dies. The cost and time savings are directly related to the fact that

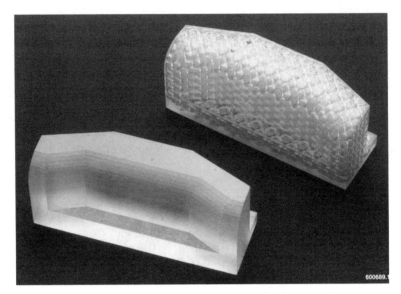

Figure 8.7   Accurate clear epoxy solid (ACES) and unfilled Quickcast parts.

complexity has no effect on the speed of creating RP parts. For example, the most complex custom louver dies were completed in 2 hr, while machining the same part required a total of 96 person-hours.

Cost savings could be increased even further by a flex press (fluid cell forming press), which conforms even better to the shape of the form block than the Guerin process rubber pad forming hydropress that we had available. The flex press shapes fine geometry such as these louvers, possibly without needing a top die. This saves considerable time and allows different thicknesses of sheet metal with the same form block.

Another unique application is the SL mold as a layup tool for composite material. The material was a fiberglass/epoxy prepreg called LTM10, which

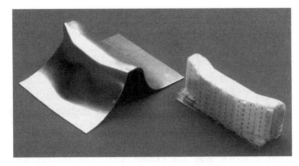

Figure 8.8   Pressed sheet-metal part and QuickCast part for initial test.

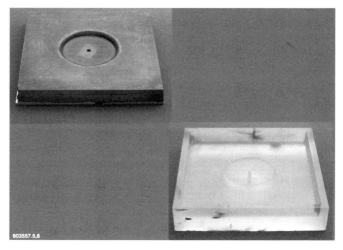

Figure 8.9   Steel-filled epoxy form block shown with mold.

cures at 65°C (150°F), allowing the high-melting-point SL5170 epoxy resin to be the tool. The composite cures for several hours in the tool and is then removed for a final cure. This was very successful, and, again, the tool required a total of approximately 4 person-hours to create the mold in stereolithography, versus roughly 60 person-hours for conventional methods.

Jim Mishek, president of Vista Technologies, extended the above work at his company. He wanted to see if forming tools could withstand the pressure and shock of high-impact conventional punching operations. Mishek used SL forming tools to punch parts with a 30-ton Amada Vipros turret punch press. The results were remarkable. Three samples of 1.5-mm (0.060-in.) thick aluminum were run, followed by five samples of 1.5-nm (0.060-in.) cold rolled mild steel, and two samples of galvanized mild steel. At first the metal was tearing, until a good radius was added. The tearing was caused by the geometry of the tool and the ductility and drawability of the material. The form blocks ran well with no measurable tool wear. When bending stainless steel, the tool began to degrade but was still within industry specifications after six parts were made!

8.4.3   Fused Deposition Modeling

An ABS plastic as a form block, built with Stratasys's fused deposition modeling FDM 1650 machine through Loyola Marymount University [16] was also tested. The compressive strength of ABS created by the FDM process is not available at this time. The same STL files that were used for the creation of the stereolithography custom louver dies became the basis for the FDM process. The dies were created in ABS plastic and put in the hydraulic press with a 1.27-mm (0.050-in.) thick, 6061-0 condition aluminum alloy sheet-metal blank. The results showed no wear at all on the ABS dies, indicating that FDM is also capable of the same cost savings as the stereolithography process when applied to sheet-metal forming.

8.4.4   Laminated Object Modeling

The same sheet-metal project was also applied to Helisys' new composite material, LGF045 [4]. Sung Pak and James Ogg, of the R&D Department at Helisys, asked to check its suitability for sheet-metal forming. Tests of the compressive strength at Helisys indicated that forming operations should work with this new material, which is made of organic fibers, an inorganic particulate, and an acrylic binder. The adhesive for bonding the layers together is a B-staged epoxy. Epoxy has three stages: A is virgin, B is partly cured (heat treated and not as tacky as A), and C is fully cured.

The material is dispensed from a roll and is 0.114 mm (0.0045 in.) thick. The same custom louver dies were grown by the LOM process and were just as successful as FDM and stereolithography. The sheet-metal blanks were 6061-0 condition aluminum alloy and were formed by a force of 10 tons. No wear was noticeable on the dies. Labor hours were 0.5 hr setup and 2 hr to do finishing. However, the 2-hr postcure is not considered labor time.

Another tool was also tested under 350 tons of force. This was the same type of tool used earlier as a steel-filled epoxy made from a stereolithography mold and is shown in Figure 8.9. The tool formed the 6061-0 condition aluminum alloy blanks with no visible tool degradation.

8.4.5   Research Results

As shown in the above research, the 98% costs savings in terms of labor hours of using stereolithography is nothing short of phenomenal [16]. The tie for second place in the 1997 North American Stereolithography Excellence Award is further proof of this success attested to by peers in the industry.

The above research shows that the same type of cost savings achieved by stereolithography can be achieved by the Stratasys process with the ABS material and the Helisys process with the LGF045 composite material. All tests came out equal to stereolithography for Helisys LGF045. However, the final tests with 350 tons of force on Stratasys ABS material has not yet been completed (only the 10-ton test).

Stereolithography resin form blocks have directly formed sheet metal in small lot production for several years at Northrop Grumman. Now that the same cost savings are being found in the other rapid prototyping technologies, it is time for more companies to transition to this new capability.

8.5   SUMMARY

In this chapter, the principles, classifications, and applications of rapid tooling (RT) have been presented. RT is a process that uses both rapid prototyping and conventional tooling processes to reduce both time and expense of product design and manufacturing.

There are generally two approaches to rapid tooling: indirect and direct. Indirect approach used RP master pattern to make molds while the direct approach uses the actual core and cavity mold inserts created by an RP machine.

There are many tooling methods available for indirect and direct rapid tooling. Although indirect methods are more established, more and more emphasis is being placed on direct method because it eliminates the intermediate step of patterning. Some of the RT methods discussed in this chapter are RTV silicone rubber mold, sprayed material, 3D Keltool, direct AIM, RapidTool, and DirectTool. Finally, a case study on sheet-metal forming using rapid tooling was presented that showed the remarkable cost-cutting ability that is available with the use of rapid tooling.

PROBLEMS

8.1. What is rapid tooling? List five advantages of RT.

8.2. What are the different types of RT? Give examples of each.

8.3. What is Keltool? Show how it works with figures.

8.4. What is meant by RTV? Describe how it works with figures.

8.5. Select three rapid tooling processes from the direct method and establish a chart comparing them in terms of lead time, relative cost, and materials and tolerances.

8.6. Select three rapid tooling processes from the indirect method and establish a chart comparing them in terms of lead time, relative cost, and materials and tolerances.

8.7. What is LENS? How does it work? What is the advantage of LENS over other tooling methods?

8.8. What is direct metal laser sintering (DMLS)? How is it different from RapidSteel form of RT?

8.9. Can investment casting be used as rapid tooling? What are the advantages and disadvantages of investment casting?

8.10. What is sheet-metal forming? Is it a convenient and economic rapid tool? How would you compare it with respect to rapid prototyping?

REFERENCES

1. *http://www.efunda.com/processes/rapid_prototyping/rt.cfm*
2. K. P. Karunakaran, P. V. Shanmuganathan, S. Roth-Koch, and K. W. Koch, Direct Rapid Prototyping of Tools, Solid Freeform Fabrication Symposium, University of Texas at Austin, 1998.
3. *http://home.att.net/~castsleisland.*
4. *http://www.wholersassociates.com.*
5. K. Lee, *Principles of CAD/CAM/CAE Systems,* Addison Wesley Longman, Reading, MA, 1999.
6. K. G. Cooper, *Rapid Prototyping Technology: Selection and Application,* Marcel Dekker, New York, 2001.
7. *http://www.armstrongmold.com/pages/rapidarticle.html.*
8. *http://www.materialise.com/prototypingsolutions/vacuum_ENG.html.*
9. *http://www.tekcast.com/sixeasy.htm.*
10. *http://www.vmreg.com/raptia/Reports/compared.pdf.*
11. *http://www.inel.gov/featurestories/01-01rsptooling.shtml.*
12. *http://129.69.86.144/raptec/outcomes/process_chains/directinjection/dmls.html.*
13. *http://www.paramountind.com/rapidp.htm.*
14. *http://www.manufacturingcenter.com/tooling/archives/1197/1197tu.html.*
15. *http://www.prometal-rt.com/process.html.*
16. B. Fritz, and R. Noorani, *Form Sheet Metal with RP Tooling, Advanced Materials & Processes,* American Society of Metals (ASM) International Materials Park, OH, 1999, pp. 37–39.

9

MEDICAL APPLICATIONS OF RAPID PROTOTYPING

One area in which rapid prototyping is having a great impact is the medical field. Some of the applications include surgical planning and the fabrication of prosthetics. The objectives of this chapter are to compile information on some current medical applications of rapid prototyping technology and to become familiar with imaging technology including new computer software packages that interface medical practices with rapid prototyping techniques. The materials for medical RP and other applications such as anthropology and biochemistry will also be discussed.

9.1   INTRODUCTION

The field of medicine is one that benefits greatly from the development and improvement of rapid prototyping technology. The number of medical applications of rapid prototyping are increasing everyday, making the future of RP more and more promising. The most obvious application is as a means to design, develop, and manufacture medical devices and instrumentation. This is an outgrowth of recognized engineering applications of the technology. Examples of medical instruments using this technology include retractors, scalpels, surgical fasteners, display systems, and many other devices [1]. However, the use of RP models extends beyond instrumentation to include surgical implants and anatomical models. The ability to accurately model parts of the human body that were only seen through 2D imaging technology affects a wide variety of fields including research and development, surgical procedures, teaching, forensics, and anthropology.

A positive aspect of RP technology is that it is capable of utilizing the existing medical imaging technology. This means that most facilities are already equipped to incorporate rapid prototyping into their existing processes. In order to further the collaboration of the medical field with RP, a variety of software packages has been developed that will take the medical image data and segment it into slices so that a 3D reconstruction can be made. Medical-grade rapid prototyping materials have also been developed, in addition to the normal RP materials, whose selection depends on the particular application of the model.

9.2 MEDICAL APPLICATIONS OF RP

The applications of rapid prototyping are not limited to only a few specific areas in the medical field. In fact, the influence of RP in the medical field is far reaching, ranging from theoretical and research areas to hands-on surgical applications. Some of the many fields of application are discussed next [2]:

9.2.1 Presurgical Planning Models

The use of RP for surgical purposes not only makes complex procedures safer but also shortens the duration of the surgery. With the aid of computer tomography (CT) scans, models of a patient's bones can be obtained. These models can then be used to rehearse complex procedures, thus reducing the actual time spent in surgery. The RP models provide the physicians and surgical team a visual aid that can be used in order to better plan a surgery so that the desired outcome is more predictable. The use of models is especially valuable in cases where surgeries are performed for the purpose of correcting anatomical abnormalities or deformities.

9.2.2 Mechanical Bone Replicas

Rapid prototyping can be used to replicate the material variations and mechanical characteristics within a bone. A lattice structure of SLA can be used to create two distinct regions that have properties similar to cortical and trabecular bones. The replicas provide mechanically correct bones, which can be used to observe the bone strength under various conditions. These replicas are more true to life because of the fact that nonhomogeneous regions can be modeled, whereas only solid homogeneous models could be obtained before. The ability to replicate bones allows doctors and researchers to recreate events that may cause fractures, stresses, and other changes in the bone.

9.2.3 Teaching Aids and Simulators

Rapid prototyping can be used to make models of any given part of the body or re-create a particular medical condition. These models are not only useful for the patient's physicians but for researchers and educators as well. The distribution of these models in kits to schools would provide a better and more hands-on illustration of the anatomy, with which students could practice certain procedures. The automobile industry already makes use of crash test dummies in order to measure safety standards in vehicles. The RP models could be used in a similar manner in medical schools, and in training courses, in order to perfect the student's skill without causing harm or discomfort to an actual person.

9.2.4 Customized Surgical Implants

Rapid prototyping models can be used for the fabrication of presurgical implants. This means that the model is used as a prototype from which the actual implant can be copied. The ability to create models allows physicians to create personalized implants for their patients instead of using standard implants. The use of RP provides an alternative to the use of standard sized implants that not only makes the implant more suitable and comfortable for the patient but also minimizes surgery time. With the current use of standardized implants, a patient is often left on the surgery table, while the implant is being customized to fit the patient. Generally, all metal custom implants are made ahead of time. However, with RP, a customized implant can be made beforehand, which would decrease the amount of time spent in surgery and consequently reduce the risk of surgical complications.

9.2.5 Prosthetics and Othotics

The field of prosthetics and othotics is one in which the benefits of rapid prototyping are most obvious. The use of RP in prosthetic or othotic devices is different from conventional methods because it starts with each individual patient's anatomy. The patient's specific alignment characteristics are included in the prototype design, which helps to reduce the number of times that a prosthetic/othotic has to be refitted. This not only makes things easier on the patient but also helps to cut down on cost. The standard method consists of first creating a plaster wrap of the residual limb in order to make a plaster cast. The casting is later filled with plaster to make a positive pattern for molding the prosthetic socket. However, this method neglects the fact that bony prominences must be accommodated by the prosthesis and that the weight-bearing regions of the socket require structural reinforcement. With the use of RP, a biomechanically correct geometry can be created the first time that would provide the patient with improved fit, comfort, and stability.

9.2.6 Anthropology

Rapid prototyping has proven to be beneficial to anthropologists because accurate replicas of delicate bones and artifacts can be made and then later used for study without causing any damage to the original finding. Once the replica has been created, molds can be made, measurements can be taken, and other tests can be performed on the replica, instead of on the artifact. In the case where there may be only one or two existing specimens, research can be done without the fear of harming or losing the rare specimen. By simply taking CT scans, an artifact can be re-created easily and accurately with an RP machine. The models that are built are useful for anthropologists to see changes in evolution that have taken place over vast periods of time.

9.2.7 Forensics

The re-creation of the crime scene is an old investigational method. Investigators working on homicide cases are especially interested in the reconstruction of the events leading to the murder. RP models could allow for the accurate prediction of the forces, implements, and other key events involved in the crime. Models of the damage inflicted by a perpetrator on a victim can also be shown in a court of law. Scenes can be re-created in the courtroom, which would help prosecutors to make their case more convincing for the jurors. In the case where a victim was struck and received severe brain damage, a replica of the damaged skull could be viewed by investigators and jurors alike. In many cases, the ability to reconstruct implements and events accurately would allow many forensics cases to be solved more quickly.

Milwaukee School of Engineering (MSOE) has an annual Research Experience for Undergraduates (REU) site that involves RP in biomedical applications. Some of the applications include rapid prototyping of bones, determining the accuracy of replicating the human skull with rapid prototyping, and rapid prototyping the lung vasculature [2].

9.3 TYPES OF MEDICAL IMAGING

Modern medical imaging technology has made a tremendous impact on patient diagnostic procedures over the last 25 years. The rapid prototyping technology that operates in a CAD environment provides many advantages for medical applications. The virtual images created by medical imaging equipment can be converted to 3D models for the medical professional to visualize different potential scenarios before an operation is performed.

Medical imaging first started with X-rays as a tool for seeing through the flesh and bones of a patient. The availability of sophisticated computers and improved digital image processing has resulted in improved imaging technology. An innovative technology in medical imaging uses traditional methods to create 3D models. These 3D models of the human body and its organs are then sliced, rotated, and analyzed depending on the area of interest [3]. The 3D models are constructed by taking slices of the desired body part and then stacking them together to form the full-scale model. Figure 9.1 shows a simplified version of the technique [4]. Both computers and conventional imaging methods are used to accomplish the image acquisition.

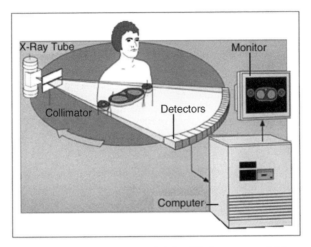

Figure 9.1   Simplified model for 3D image acquisition.

Classical imaging techniques are identified through the two-dimensional images they produce. There are several types of conventional medical imaging techniques, such as X-rays, computed tomography (CT), magnetic resonance imaging (MRI), and positron emission tomography (PET). In this section we shall briefly describe the operating principles of these techniques.

9.3.1   X-Ray Technology

X-rays are used everyday worldwide by medical professionals to look at bones that may be broken or contain stress fractures, to see something inside the body that might have been swallowed, to see images of gunshot wounds, to view gallstones, or even to scan baggage at the airport [4].

X-rays are penetrating electromagnetic radiations, having a shorter wavelength than light, and are usually produced by bombarding a target with high-speed electrons. X-rays were discovered accidentally in 1895 by the German physicist Wilhelm Conrad Roentgen while he was studying cathode rays in a high-voltage gaseous discharge tube. Despite the fact that the tube was enclosed in a black cardboard box, Roentgen observed that a barium–platinocyanide screen, inadvertently lying nearby, emitted fluorescent light whenever the tube was in operation. Upon further investigation, he determined that the fluorescent light was caused by invisible radiation of a more penetrating nature than ultraviolet rays [5]. He named the invisible radiation "X-ray" because of the unknown nature of the radiation. In Figure 9.2 an X-ray image of a human chest is shown [6].

Nature of X-Rays   X-rays are electromagnetic radiations whose wavelengths range from 100 to 0.01 Å (1 Å is about 10^{-8} cm). The shorter the wavelength of the X-ray, the greater is its energy and its penetrating power [4]. X-rays

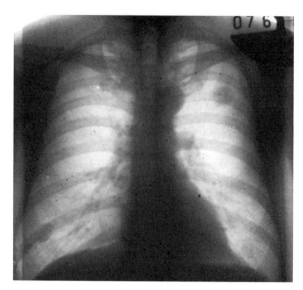

Figure 9.2 X-ray of human chest cavity.

are produced whenever high-voltage electrons strike a material object. While most of the energy of the electrons is lost in heat, the remaining energy produces X-rays by causing changes in the object's atoms as a result of the impact.

The X-rays affect a photographic emulsion in the same way as light does. Absorption of X radiation by any substance depends on its density and atomic weight. The lower the atomic weight of the material, the more transparent it is to X-rays of a given wavelength. For example, when the human body is X-rayed, the bones consisting of higher atomic weights than the surrounding flesh absorb radiation more effectively and cast dark shadows on a photographic plate. Neutron radiation, which is different from X radiation, is now also used in some types of neutron radiography, which produces the almost exact opposite results. Objects that cast dark shadows in an X-ray picture are almost always light in a neutron radiograph.

Applications of X-Rays Research has shown that X-ray technology has played a very vital role in theoretical physics, especially in quantum physics. Using X-rays as a research tool, scientists have been able to confirm experimentally the theories of crystallography. The X-ray diffraction method has been used to identify crystalline substances and to determine their structures. Virtually all modern crystalline substances were either discovered or verified by X-ray diffraction analysis. By using X-ray and diffraction methods, chemical compounds can be identified and the size of the ultramicroscopic particles can be established. Chemical elements and their isotopes are regularly identified by X-ray spectroscopy.

Some recent application of X-rays involve microradiography and stereo-radiogram. Microradiography produces fine-grain images that can be enlarged considerably. Two radiographs can be combined in a projector to produce a 3D image called a stereoradiogram. Electron microprobes also provide extremely detailed and analytical information of specimens.

In addition to its application in research, X-rays are being used by industry as a tool for testing objects such as metallic castings without destroying them. Many industrial products are inspected routinely using X-rays to find defects in the products. Ultrasoft X-rays are used to determine the authenticity of art work and for art restoration purposes.

The impact of the application of X-rays on health care is tremendous. X-ray photography, known as radiography, is used extensively in medicine everyday as a diagnostic tool. People also undergo radiotherapy where X-rays are used to treat and cure conditions such as shrinking tumors and destroying various forms of cancer. Tuberculosis was single-handedly diagnosed by means of radiation when the disease was prevalent. Modern X-ray devices can offer clear views of most of the human anatomy.

9.3.2 Magnetic Resonance Imaging

Magnetic resonance imagery, or MRI for short, was previously known as nuclear magnetic resonance imagery (NMRI). The word "nuclear" was dropped about 16 years ago to remove fear of nuclear radiation from minds of the patients, especially since the risk of harmful radiation was actually much less with MRI than with conventional X-rays [7].

The MRI technique takes pictures of various parts of the body without the use of X-rays. An MRI is safe for most patients. People who are claustro-phobic and who have implanted medical devices such as an eurysim clip in the brain, heart pacemakers, and cochlear implants may not be able to have an MRI. An MRI scanner employs a large and very strong magnet that envelopes the patient. A radio wave antenna is used to send "radio wave signals" to the body and then receive the signals back [7]. These returning signals are converted into pictures by a computer attached to the scanner. MRI is a powerful and versatile tool that generates thin-section images of any part of the body including the heart, arteries, and veins, from any angle and any direction without surgical intervention. MRIs can also be used to create "maps" of biochemical compounds within any cross section of the body. These maps provide valuable biomedical and anatomical information for new knowledge and for early diagnosis of many diseases.

How MRI Works   The human body consists of billions of atoms, the building blocks of all matter. There are many types of atoms in the body, but for the purpose of MRI, we are concerned only with the hydrogen atom. It is an ideal atom for MRI because its nucleus has only a single proton and a large magnetic moment. It means that when placed in a magnetic field, the hydrogen atom has a strong tendency to line up with the direction of the magnetic field.

The principles of MRI take advantage of the magnetic nature of these protons. Once the patient is placed inside the cylindrical magnet, the diagnostic process begins. The process contains five basic steps: (1) The MRI creates a steady state within the body by subjecting the body in a steady magnetic field that is 30,000 times stronger than Earth's magnetic field. (2) A radio frequency (RF) signal is beamed into the magnetic field, which changes the steady-state orientation of the protons. (3) When the RF signal stops, the protons move back to their previously aligned position and release energy. (4) A receiver coil measures the energy released by the disturbed protons. These measurements provide information about the type of tissue in which the protons lie as well as their condition. (5) A computer uses this information to construct an image on a TV screen.

In current medical practices, MRI is primarily used for diagnosing most diseases of the brain and the central nervous system. MRI has the special feature of distinguishing soft tissue in both normal and diseased states. Figure 9.3*a* shows a person getting an MRI scan, while Figure 9.3*b* shows an image that is obtained from the scan.

9.3.3   Computed Tomography

Computed tomography and computed axial tomography (CAT) scans are medical imaging technology that use both X-rays and computers to produce 3D data of the human body. CT scanners are relatively inexpensive and are owned by most hospitals. While X-rays traditionally highlight dense body parts, such as bones, CT scanners provide detailed images of the body's soft tissues, such as muscle tissues, blood vessels, and even organs such as the brain. Unlike conventional X-rays, which provide flat 2D images, CT provides a cross section of the body.

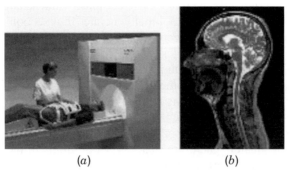

(*a*)   (*b*)

Figure 9.3   Computed tomography (CT).

A patient undergoing a CT scan is placed on a movable table at the center of a donut-shaped scanner, the size of which is about 2.4 m (8 ft) tall as shown in Figure 9.4 [8]. The CT scan system consists of an X-ray source, which emits beams of X-rays, an X-ray detector, which monitors the number of X-rays that strike the various parts of its surface, and a computer to analyze and differentiate the detected energy. The X-ray source and detector face each other on the inside of the scanner ring and are mounted so that they rotate around the rim of the scanner. X-ray beams pass through the patient and are recorded on the other side of the detector. As both the source and the detector are in a 360° circle around the patient, X-ray emissions are recorded from many angles. By moving the patients within the scanner, the operation can obtain a series of images called "slices." Each slice represents a slab of the patient's body with a certain thickness (typically 1 to 10 mm). For CT canners, each slice has 512×512 pixels. Doctors analyze these series of slices to understand the 3D structure of the body.

In order to sharpen an image, patients are sometimes injected with a substance that increases the contrast between different tissues. Patients are also asked to drink a liquid that makes the internal organs more visible in the CT scan. For scanning special parts of the body, such as the chest, abdomen, and pelvis, contrast agents are regularly used.

Since its discovery in the late 1960s and early 1970s by British engineer Sir Godfrey Newbold Hounsefield and American physicist Allan Macleod Cormack, the speed at which CT scans are obtained has increased greatly. While the old scanners required 4.5 min of scanning and 1.5 min of computer

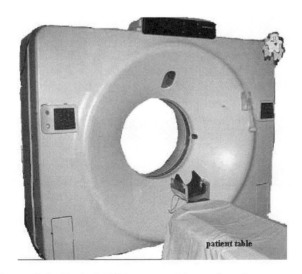

Figure 9.4 Typical CAT scan machine and equipment setup.

reconstruction, the new system of CT scanners can create a slice in about a second.

The most frequent application of CT is for viewing the tissues of the brain. CT scans of a brain can show if an accident victim has sustained brain injury or if bleeding has occurred in the brain of a stroke victim.

9.4 SOFTWARE FOR MAKING MEDICAL MODELS

The process of rapid prototyping enables us to generate a physical model from CAD geometry. The model is made by one of two methods: either a CAD program of the desired object is created first or a physical object is scanned, and a 3D CAD model is obtained through the reverse engineering process. Because the medical field is an area that greatly benefits from rapid prototyping, various companies have produced software that can be used to convert medical scan data to mathematical representations that are usable by RP or CAD systems.

9.4.1 MIMICS and the CT Modeler System

Materialise's Interactive Medical Image Control System (MIMICS) is a software system that interfaces between scanner data including CT, MRI, and technical scanners, to RP, STL file format, CAD systems, and finite element analysis (FEA) [9]. The MIMICS software is an image-processing package with 3D visualization functions that allows CT images as well as MRI images to be segmented, so that they can later become useful physical models.

The MIMICS process can be used for the purposes of conducting diagnostics, planning surgical procedures, or for teaching and rehearsal purposes. A flexible prototyping system is also included for the building of distinctive segmentation objects. MIMICS enables surgeons and radiologists to control and correct the segmentation of CT scans and MRI scans. What this means is that the object to be visualized or produced can be defined by the medical staff without any technical knowledge about creating an on-screen 3D visualization of a medical object. Separate software is available that helps to define and calculate the necessary data to build the medical objects created with MIMICS on all rapid prototyping systems. The MIMICS software is capable of processing any number of 2D images slices, with the user's computer memory serving as the only restriction.

Visualization Tools In MIMICS, segmentation masks are used to highlight regions of interest. Images can be processed and defined with up to 16 colored segmentation masks. Thresholding is the first action performed to create a segmentation mask. A region of interest (ROI) is defined by a range of gray values, with the boundaries of the threshold being the upper and lower range. This is particularly useful in segmenting bone structures within medical CT image data. Because it is possible to define the segmentation region by two thresholds, this technique can be used for the segmentation of soft tissue in CT images or for the segmentation of multiple structures in MRI images. Additional functions such as the profile function and the histogram function are available to help define distinct thresholds when objects with different materials are scanned. Figure 9.5 shows a human hip along with the pelvis, spine, and femurs, shown on a MIMICS screen [9]. The figure also shows the multiple screens that are available, including a display of a 2D slice, the CT image, and the 3D reconstruction of the pelvic region.

Visualization and Measurements Tools MIMICS displays CT and MRI image data in several ways, each of which provides unique information. MIMICS divides the screen into three views: the original axial view and the resliced data making up the coronal and sagittal views. In addition, various options are provided such as a zoom factor from 1 to 120, contrast enhancement, as well as the ability to pan across the view and rotate the 3D calculated data. Also provided is a flexible interface to visualize the segmented objects as 3D objects using fast and advanced 3D rendering and shading algorithms.

Measurements can be taken within MIMICS from point to point on 2D slices as well as 3D reconstructions. A profile line displays an intensity profile of the gray values along a user-defined line. Accurate measurements can be made based on the gray value by three methods: the threshold method, the

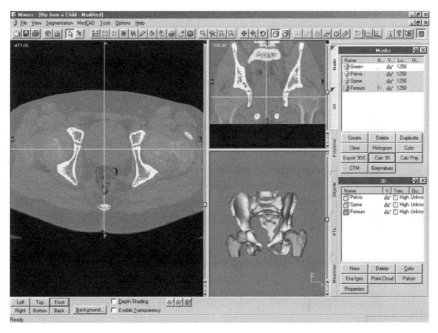

Figure 9.5 Example of MIMICS image with multiple threshold masks. (Courtesy of Materialise, Inc.)

four-point method, and the four-interval method, all of which are ideal for technical CT users.

9.4.2 Velocity²Pro Software by Image3

Velocity²Pro is a software for Windows NT and Silicon Graphics workstation that allows for the production of 3D reconstructions directly from CT and MRI image data using gray level isosurfacing and mask definitions. Complex parts such as vessel structures, floating tissues, bone splinters, and internal cavities can be generated seamlessly throughout the entire model surface.

Every Velocity²Pro module opens as one or more windows that can easily be moved or resized. Functions are icon based, which makes work fast and user friendly. The Velocity²Pro comprehensive image processing allows for the restoration of low-contrast images, the masking out of artifacts, thresholding, drawing, painting, filling, and the automation of data using scripts. Similar to MIMICS software, Velocity²Pro's gray-level operations include area histograms, scaling, and binary-level slicing to name a few.

Visualization and Measurement Velocity²Pro's Display module allows the user to work with multiple assemblies at once, utilize multiple colored light sources, and to create solid-translucent assemblies from the same data set. For instance, a model of a person's head can be made in which the brain can be seen right through the face. Figure 9.6 shows a reconstruction of CT image data of a young patient with congenital hip dysplasia (approx. 50 slices). The data is courtesy of Shriner's Hospital for Children [10]. Measurement and analysis tools are available for 3D surfaces and volumes. Many functions such

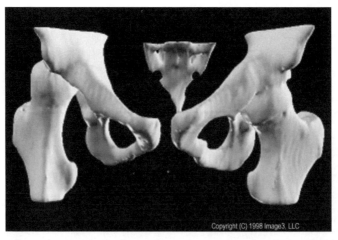

Copyright (C) 1998 Image3, LLC

Figure 9.6 Shows a 3D rendering of a pelvis using Velocity²Pro.

as viewing, rotating, and capturing of screens are similar to those of widely used solid modeling software such as Pro-E and SolidWorks.

9.4.3 3D-Doctor from Able Software

3D-Doctor is an advanced 3D image visualization, rendering, and measurement software for MRI, CT, microscopy, scientific, and industrial 3D imaging application, to be used on a PC, that has also earned the distinction of being U.S. FDA approved [11]. 3D-Doctor is capable of creating 3D surface models or volumetric renderings from 2D cross-section imaging, and exporting the 3D models to DXF, 3DS, STL, IGES, VRML, and other file formats. It supports most commonly used 3D and 2D image file formats including BTM and JPEG, and can scan CT/MRI films using a regular image scanner and 3D-Doctor's template-based crop film function to separate the slices with a few simple clicks of the mouse. Multiple objects can be combined to create 3D views of complex structures for applications such as surgical planning and diagnostic imaging. Parameters such as material properties, colors, viewing angles, and display settings can be changed at will, which makes the viewing of the 3D rendering easier and more informative.

Visualization and Measurement The visualization functions offered by 3D-Doctor allows one to view a 3D image in two different windows, the 2D image slice in a third window, and a montage window, where all of the image planes are arranged in an array for 2D image analysis. This means that if you are not happy with the slice you are analyzing, you could pick the correct slice, and automatically get the 2D and 3D images that you need. Figure 9.7 shows the multiple windows that are offered by 3D-Doctor [11]. The palette control function allows the user to change the window display to pseudocolor, red, blue, green, or gray scale. The image can be adjusted for contrast or pixel interpolation by the user.

3D-Doctor's measurement tools allow the user to measure the length and area of a region, 3D surface area, volume, image density in a region, and the pixel histogram, all of which are useful for analysis. A reslice command is offered so that the CT/MRI image can also be resliced into smaller user-defined measurements.

9.4.4 VoXim

VoXim is a software that was designed specifically for the Microsoft Windows operating systems by IVS Solutions [12]. With VoXim, three-dimensional diagnosis and therapy planning in a wide variety of fields can be done quickly and efficiently. VoXim, like all of the other software packages discussed, is able to segment CT and MRI image data. Unlike most other systems, VoXim is completely volume based, which provides the user with better quality 3D

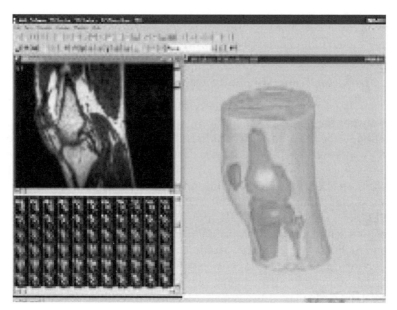

Figure 9.7 Multiple windows offered by 3D-Doctor. (Courtesy of Able Software Corp.)

images. The VoXim/Modeling module provides segmented objects to be used for CAD or RP; therefore, it supports standardized data types such as CLI, DXF, IGES, and STL. The VoXim/Modeling module also supports parameters such as enhancement, scaling, offsetting, and mirroring to name a few. Other enhancement modules that are offered with VoXim are NeuroMate, Stereotaxy, Osteotomy, Animator, and Conferencing. Figure 9.8 shows a 3D image of a skull provided by VoXim [12].

9.5 MEDICAL MATERIALS

The RP models for use with medical applications can be made with a variety of materials. In some cases though, these materials used must be "medical grade." Factors that determine what materials to use for an application are much the same factors that one must take into consideration when designing any industrial application. For medical use, practice and simulation models have few constraints as to what kind of material should be used, as the sole purpose of these models is fit check and geometry verification. However, in some cases RP models are used as surgical tools or may be actually implanted to permanently remain in vivo. These types of applications require models with the ability to be sterilized or remain compatible with human tissue. Unfortunately, materials that can be permanently implanted are still in developmental phases. Table 9.1 shows different RP methods, materials used in

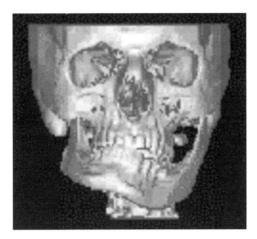

Figure 9.8 Skull image provided by VoXim.

these methods, and processes of sterilization [13]. These models are not currently approved for permanent implants.

Models produced in the Sterocol material from Vantico can include colored regions of the anatomy to highlight a particular feature. This is the only stereolithography resin that allows for a demarcation of a radically different second color within the model. This feature can be used for more than highlighting a particular feature. It can also be useful to create markings on implants to help the surgeon know exactly where certain tasks are to be completed. For example, the path of a screw through an implant or practice model can be highlighted. Figure 9.9 shows the teeth and their roots in a model of a jaw [9].

Even though RP materials are still in the developmental stages for medical applications, the benefits are quite apparent. As technology expands, one of the research goals may be to directly copy a patient's anatomical characteristics, such as to fabricate materials that have strengths equivalent to human bones.

TABLE 9.1   RP Materials for Medical Use

RP Method	Material (Vendor Name)	Method of Sterilization
Stereolithography	Sterocol (Vantico)	Autoclave, γ irradiation, ethylene oxide
Selective laser sintering	DuraForm polyamide (3D Systems)	Autoclave
Fused deposition modeling	Medical-grade ABS (Stratasys)	γ irradiation

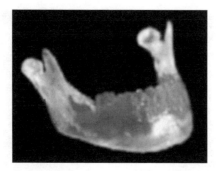

Figure 9.9   Example of color stereolithography.

9.6   OTHER APPLICATIONS

While the connection between rapid prototyping and the medical industry are quite evident, there are nevertheless other research areas in which the use of RP could prove to be invaluable. Other such categories include anthropology, biology, and chemistry. Although these three fields do not directly aid doctors and other medical professionals with the treatment of patients, continued research in these areas could lead to many medical breakthroughs.

9.6.1   Anthropology

The fields of anthropology and paleontology rely almost entirely on very old and very fragile artifacts; therefore, the greatest of care is required when handling the delicate fossil remains. Oftentimes, the study and research of artifacts are hindered because of the fear of damage, not to mention the limitations on the ability for the public to view these scientific breakthroughs. However, with the development of three-dimensional medical imaging and rapid prototyping, the artifacts can be scanned and reproduced with such a degree of accuracy that the prototyped objects can be used for both research and viewing.

For instance, the renowned Smithsonian Institution's National Museum of Natural History is currently using RP technology to re-create and replace a triceratops skeleton that has been standing since 1905 [14]. Because the original skeleton is at least 65 million years old, and has been standing on its feet for the last 100 years, it has been steadily weakening and is in danger of collapsing. Therefore, a preservation committee has taken hundreds of scans in an effort to replicate the skeleton. Initially, in 1905, no actual triceratops feet were found; therefore, feet from another dinosaur were used to erect the model. As a result of the more current information, and the ability to replicate bones that were initially missing, the new triceratops skeleton will be more complete, more accurate, and stand taller than the original skeleton

[14]. The remarkable resemblance between the actual skeleton and the re-created one is shown in Figures 9.10*a* and *b*.

9.6.2 Biology and Chemistry

Advancements in microbiology and chemistry are crucial to improvements made in the medical field. The ability to visualize components at the microscopic level is a key that will allow scientists and researchers to better understand the workings of some of the smallest yet most important features in the human body. For instance, at MSOE, various researchers are using rapid prototyping as a tool for producing models of microscopic cellular details such as proteins, deoxyribonucleic acid (DNA), and enzymes [2].

Companies such as 3D Molecular Designs specialize in creating physical models of proteins and other molecular structures using rapid prototyping. This startup company was developed at the Center for BioMolecular Modeling at MSOE and utilizes the university's RP facilities. A model of a nucleosome is shown in Figure 9.11.

9.7 SUMMARY

In this chapter, we have presented the many applications of rapid prototyping in the medical field. The chapter provides details of how and why rapid prototyping is being applied in varying applications such as surgical planning, implant and prosthetic modeling, medical bone replicas, and teaching aid simulators.

The various types of medical imaging technology and their principles are discussed, and the link between medical imaging and RP has been established. The conventional medical imaging techniques discussed in detail were X-rays, computed tomography (CT), and magnetic resonance imaging (MRI). Spe-

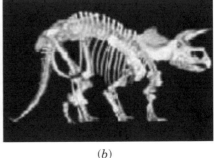

(*a*)   (*b*)

Figure 9.10   (*a*) Original triceratops skeleton and (*b*) computer rendering of original skeleton.

Figure 9.11 Nucleosome model made from 3D printing.

cialized software packages have been developed to interpret the medical image data and format it correctly for use with various RP machines and techniques. Therefore, with the information obtained from medical imaging, and the correct software, a 3D model can be built for any region of the body. Some of the software programs are Materialise's Interactive Medical Imaging Control System (MIMICS), Velocity^{2}Pro, 3D Doctor, and VoXim. All of these programs allow the user to view the 2D image from a CT scan or an MRI and then render the desired part or parts in 3D.

The applications of RP are not limited only to hospitals; many research labs also benefit from the use of rapid prototyping including fields such as anthropology and biochemistry. The ability to safely scan and model delicate fossils and view components such as proteins and enzymes that are at the microscopic level can greatly help scientists and researchers unlock the mysteries of the human body at the most basic cellular level. The usefulness of RP in the medical field is increasing everyday and is now one of the most promising areas of rapid prototyping.

PROBLEMS

9.1. Name five areas of application for rapid prototyping in the medical field and briefly describe how RP technology is improving each area.

9.2. What is an X-ray? How does it work? What are some of the nonmedical uses of X-rays?

9.3. How is an MRI different from an X-ray? What are the five basic steps used in order to obtain an MRI image?

9.4. How is a CT scan an improvement over a regular X-ray scan?

9.5. Name three of the medical software discussed in the chapter and discuss the benefits of each.

9.6. If you were a surgeon about to perform a delicate operation to remove a tumor in a patient's brain, which software would you use and why?

9.7. What are some of the considerations and constraints that need to be taken into account while developing a medical-grade material?

9.8. How would the development of a medical-grade material with properties similar to human bones be better than the standard titanium implants used today?

REFERENCES

1. J. Alcisto and C. Parrish, *General Applications of Rapid Prototyping,* Loyola Marymount University, Los Angeles, CA, 2001.

2. Milwaukee School of Engineering (MSOE), Rapid Prototyping Center, *www.rpc.msoe.edu/3dmd/3dmd.php, 2002.*

3. J. Jankowski and M. La Mothe, *Medical Application in Rapid Prototyping,* Loyola Marymount University, Los Angeles, CA, 2001.

4. X-ary, *http://www.cfi.lbl.gov/~budinger/medTechdocs/Xray/html.*

5. Microsoft Encarta, Redmond, WA, 2001.

6. The Virtual Human Body—State of the Art and Visions for Medicine International Symposium 1, Hamburg, Germany, 2002, *http://www.uke.unihamburg.de/institute/imdm/idv/konferenzen/symposium2002/abstracts/toriwaki.html.*

7. Kaiser Permanente, *Magnetic Resonance Imaging—An Inside Look,* Los Angeles, CA, 2002.

8. *http://www3.lbcc.cc.ca.us/coursepages/eng3sw/hypertextsfall99/imaging/intro1.html.*

9. Materialise, Belgium, *http://www.materialise.com.*

10. Image3, LLC—Velocity²Pro, Utah, *http://www.image3.com.*

11. Able Software Corp., Massachusetts, *http://www.ablesw.com.*

12. IVS Solutions AG—Voxim, Germany, *http://www.voxim.com.*

13. T. Wohler, *Wohler's Report 2002,* Fort Collins, CO, 2002.

14. *Relief for Weary Bones. www.graduatingengineer.com;* article: *www.asme.org,* 2002.

10

INDUSTRY PERSPECTIVES

The ever-increasing demand for quality products and lower costs has manufacturers turning to the new technology of rapid prototyping. The successful implementation of rapid prototyping depends on the basic principle of planning. This chapter discusses the guidelines for implementing RP and managing issues related to this new technology. There are alternatives to purchasing an RP system. Service bureaus and consortia are organizations that can provide the RP needs of companies. This chapter also describes what they are, how they operate, and some criteria for selecting a service bureau that will provide easy and efficient access to this versatile new technology.

10.1 GUIDELINES FOR IMPLEMENTATION

There is no denying the fact that commercialized rapid prototyping systems have had a profound impact on product development around the world. However, purchasing a rapid prototyping system and implementing it in the workplace can be expensive and fraught with unexpected pitfalls. In this section, we shall discuss guidelines for purchasing decisions, facilities planning, and environmental issues.

10.1.1 Purchasing Decisions

Before purchasing a rapid prototyping system, two practical but commonsense questions should be considered. First, should a particular rapid prototyping process be selected over other prototyping process for a given type of product? Second, which rapid prototyping process should be chosen based on which aspect of the technology will be pursued by this company? A number of criteria should be evaluated when considering the first question. These criteria depend on the product application and may include [1]:

- Cost trade-off
- Cycle time (throughput)
- Accuracy of the prototyped parts including tolerance and surface finish
- Material properties (material selection)
- Size of the part
- Strength of the part

These criteria should be carefully considered and traded-off between competing processes. Sound decisions should be made based on sound business practice and economics. Although the ultimate factor depends on the particular product being developed, questions such as in-house model making versus outsourcing prototypes should be considered. If in-house model making is not available, out-sourcing with service providers should be considered. As a first-time user, outsourcing is always recommended in order to become familiar with various processes and the material properties of the resultant prototypes.

The second question of which prototyping process is preferred also depends on the above criteria. Additional factors such as the complexity of the system, size, weight, and operational costs, operating environment (office versus shop floor), material waste, and availability of CAD data and personnel (training and operating) requirements should also be evaluated. Maintenance and costs related to service contracts can also be a significant factor in making a purchase decision.

10.1.2   Facilities Planning

Facilities planning generally deal with the planning and design of location as well as the building and equipment of an organization. Once the decision is made to purchase a rapid prototyping system, it is necessary to do some facility planning before installing the system. For example, most stereolithography-based systems require the facilities to have proper ventilation. The facility must have the capability to cycle the room air at least six times a minute. Postcuring and cleaning equipment also will likely require ventilation as well as the necessary storage required for solvents and other hazardous materials. As for electrical needs, each system has its own unique requirements, but a dedicated and spike-free circuit should always be planned for this type of equipment. There should be adequate space for the prototyping systems as well as the cleaning and support equipment. A rapid prototyping facility should be located inside a plant in such a way that is easily accessible to customers, suppliers, skilled labor pool, and design/manufacturing engineers. A recurring commitment to familiarizing technical personnel to the uses for advantages of RP will improve utilization and benefits of the system.

10.1.3   Environmental Issues

Some RP systems require no special facilities or venting and involve no hazardous materials or by-products. Stratasys' RP machines are excellent examples of an environmentally friendly systems. The new Stratasys Titan is presented in Figure 10.1, and it shows an example of a machine that needs no added ventilation and minimal support equipment.

There are many RP machines that are not as environmentally friendly. These machines usually require a specialist to come to the site where the machine is to be set up and certify that the system meets the Environmental Protection Agency (EPA) standards. Some examples of machines that are environmentally sensitive include selective laser sintering (SLS) machines, SLA machines that use resins, and laminated object manufacturing (LOM) machines such as the one pictured in Figure 10.2.

Figure 10.1   Stratasys' Titan system shows no ventilation.

Figure 10.2 LOM with added ventilation.

10.2 OPERATING ISSUES

10.2.1 Cost Analysis

There are many factors that must be examined when considering the purchase of a rapid prototyping system. Cost analysis is a huge factor in this process. Included in the cost analysis are the land and building costs, the primary equipment costs, and optional equipment costs. Facility modification costs, operating expenses, and other miscellaneous expenses are also included when considering the purchase of a rapid prototyping system. The following briefly describes what is involved in these costs. One must take into consideration that the following data was taken in the early 1990s, so the costs should be adjusted as required to suit the time and location of the proposed installation.

Land and building costs are important in any operation. Presuming that a building already exists for this equipment, the appropriate floor space must be allocated as close to the primary users as possible. The cost per square foot should be considered the same as laboratory space.

Today, rapid prototyping systems (excluding the CAD system) range in cost from $30,000 to about $700,000. Some of the developers are considering the introduction of systems for under $30,000 in the future, which will be designed for use in an office environment. Table 10.1 is a partial list of current (2005) prices of available rapid prototyping machines. Total equipment costs include machine, material, accessories, and post-processing equipment (if applicable), installation, and training.

Additional equipment may include a workstation and a CAD/CAM package such as Pro-Engineer. Some examples of this equipment and its respective costs are presented in Table 10.2. As can be seen in Table 10.2, it is a good idea to budget between $25,000 and $40,000 for one CAD system, complete with software and hardware. Several days of training and learning, combined with weeks of hands-on practice should also be considered.

TABLE 10.1   Approximate Cost of RP Equipment in 2004

RP Systems	Estimated Price
3D System's Viper SLA	$320,000
3D System's InVision	$40,000
3D System's Sinterstatin® HiQ™	$450,000
EOS's EOSINP P 380	$430,000
Fused Deposition Modeling (FDM) FDM-Prodigy Plus	$55,000
Fused Deposition Modeling (FDM) FDM-Titan	$190,000
Solid Object Ultra-Violet Laser Plotting (SOUP) 1000GA	$600,000
Z-Corp's Spectrum Z510	$55,400
Z-Corporation 3D Printer (ZPrinter 310 Station)	$28,800
Object's Eden 260	$90,000
Object's Eden 333	$115,000

Facility modification costs include any kind of changes that might need to be made to accommodate the new RP machine and equipment. For instance, wall partitions and ceilings might need to be added to separate the machine from the other work environments around it. Because certain materials used by the RP machines, such as resins, are toxic, the company may require proper ventilation to be added. Generally, the room must have the capacity to change the room air at least six times a minute. There needs to be adequate space for the RP unit, CAD system, postcure ovens (if required), cleaning equipment, finishing area, and storage of supplies. Figure 10.3 shows 3D Systems' Sinterstation 2500 as an example of one of the larger systems.

One of the major operating expenses of some systems is the cost of raw materials. Raw material may be liquid resin in the case of SLA, a roll of paper for LOM, or a roll of filament for FDM. A mile-long spool of wax or nylon filament currently (2004) costs approximately $200, which is equivalent to 1 gal of polymer; 25 to 30 spools are probably used per year, which makes the cost of materials between $5000 and $6000 per year. Another example of material cost is the SLA 500. The SLA 500 has a 67-gal vat that is going to cost $13,400 to fill for the first time. The cost of materials is usually not included in the initial purchase price of the equipment.

TABLE 10.2   Cost of Optional Equipment, 2005

CAD/CAM	Estimated Price
Software	Pro-Engineer Advantage software: $5000/seat
Training	Basic: $1500 (3 days)
Maintenance	$1500
Hardware	Pentium PC: $2,000
	SGI workstation: $5,000

Figure 10.3 3D Systems' Sinterstation 2500.

The salaries of the personnel also fall under the operating expenses. Operating personnel for most RP systems should include one engineer and one technician for efficient utilization and management of the system.

Another major component of the operating expenses is the maintenance cost. Maintenance agreements for RP systems can cost between $16,000 and $36,000 or more. The actual costs for the agreements vary due to the fact that some machines are more or less complex than others. For example, 3D Systems currently charges $22k–$32k per year for the SLS maintenance agreement. This may or may not include periodic replacement of the laser. In general, RP machines without helium cadmium or argon lasers will cost less to maintain.

Other miscellaneous costs include all the pre- and postprocessing equipment needed to prepare and finish the object. For some machines a postcure oven is needed. Ultrasonic cleaners and other cleaning tools are needed for most RP machines. Protective equipment and material storage facilities should also be considered as part of miscellaneous operating expenses.

10.2.2 Training of Personnel

Selecting the right rapid prototyping system is absolutely essential for any company that has identified a need for this equipment. However, the success of the implementation of the rapid prototyping system depends largely on the people of the company who operate and maintain the rapid prototyping machine and its support equipment [2].

There is a myth about rapid prototyping machines that "the systems are as simple as pushing a button." Not so. As with any complex system, it can take a company several weeks to more than a month to get the machine running properly. Based on a positive experience with their first machine, many companies have bought several additional machines for higher productivity and reduced cost and waste. Rapid prototyping is a new technology, and as such, it takes a great deal of experimentation and practice to make quality prototypes that will be useful to the end users.

When a machine is purchased, the machine vendor should be contracted to train the people responsible for the operation and maintenance of the machine. This training is critical for the successful implementation of a prototyping machine into the work environment. Depending on volume of parts produced, it may be advantageous to dedicate one trained person exclusively for part finishing. Most parts for finishing equipment can be purchased from either the machine vendor or from other sources of industrial equipment. The cost of finishing equipment will depend on quality required of the finished prototypes.

10.2.3   Maintenance Costs

Maintenance of rapid prototyping system is absolutely essential for the long-lasting, trouble-free operation of the machine. System maintenance varies from system to system. The first thing to investigate is whether there is a local representative to service the system. A local representative can save travel time and expense in servicing the machine, and most system vendors offer maintenance agreements at the time of sale. For companies that use RP machines regularly, it is advisable to have a service agreement with the vendor.

The cost for systems that use lasers as energy sources, such as SLA 250 or DTM Sinterstation, is generally higher. For example, the cost for yearly service agreement for SLA 250 from 3D Systems can be around $30,000 to $40,000. On the other hand, Stratasys charges only $15,000 for the yearly maintenance agreement for its FDM 3000.

Many small companies and universities cannot afford to have such agreements because of high cost. In that case, an experienced operator who can operate, maintain, and troubleshoot the machine is a plus.

10.3   MANAGING ISSUES

10.3.1   Integrating Technology Within the Company

As mentioned in Chapter 1, rapid prototyping is not a technology in itself; rather it draws on other technologies such as CAD/CAM, CNC machining, and the like. RP has made significant contributions to the design and manufacturing industries mainly by reducing the time to produce prototype parts and improving the ability to visualize part geometry. Integrating the RP technology into the daily activities of any company can be critical to the success of the company. The full production of any new product involves a wide spectrum of activity. The impact and integration of rapid prototyping technologies for product development and process realization [2] is shown in Figure 10.4. With reference to Figure 10.4, the activities for full production in a conventional model are shown at the top section of the figure. The impact of RP technology over the past 30 years is shown at the bottom. Once RP technology is integrated into the company, savings in time and cost could range from 50 up to 90% [3].

10.3.2 Productivity Advantage

Today's advanced RP systems can produce functional parts in limited production quantities. Parts thus produced have an accuracy and surface finish slightly inferior to those made by conventional NC machines. However, the RP systems can produce prototype parts faster than any conventional machining system. The benefits of RP systems are generally classified into *direct* and *indirect* benefits. The productivity benefits are considered direct benefits.

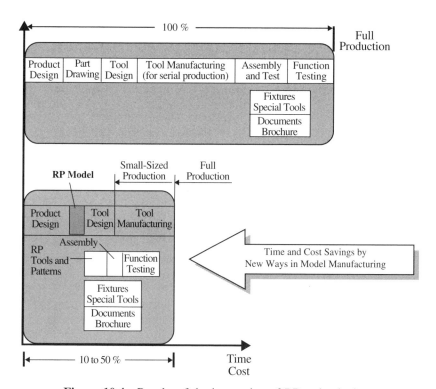

Figure 10.4 Results of the integration of RP technologies.

Direct Benefits   There are many direct benefits of RP to a company. One such benefit is the ability to experiment with complex physical objects in a relatively short period of time. Over the last 25 to 30 years, the complexity of products has increased dramatically. For example, the aerodynamics and aesthetics used in today's design of car bodies are very much different from those of the 1970s. It is seen in Figure 10.5 that the factor of complexity has increased from 1 in the 1970s to about 3 in the 1990s. However, the time for the completion of a project has not increased substantially. Figure 10.5 shows a modest increase in project completion from 4 weeks in 1970 to 16 weeks in 1980. However, with the utilization of CAD/CAM and CNC technologies, project completion times are reduced to 8 weeks [2]. The availability and use of RP further reduced the project completion times to about 3 weeks as of 1995.

For procurement departments, the availability of prototypes can greatly enhance the fidelity and accuracy of vendor bids, especially on complex products with hidden features that may not be obvious in the CAD model.

Benefits to Product and Tool Designers   Product designers receive many direct benefits from RP. The product designers can use the RP model to increase part complexity without sacrificing the lead time and cost. The designer can optimize a part design to meet customer requirements with minimal manufacturing restrictions. The number of parts in an assembly can be reduced by combining several features into one piece. This leads to less time spent on tolerance analysis, fastener selection, and so forth.

Tool designers can also eliminate costly and time-consuming practices for machining metal tools and dies. The additive nature of rapid prototyping is helping companies supplement their traditional process with rapid tooling to

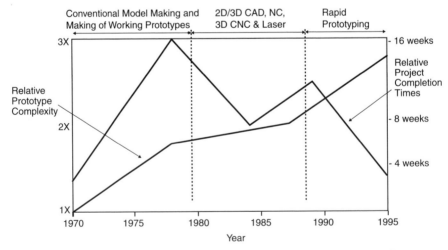

Figure 10.5   Project time and product complexity in a 25-year time frame.

produce parts in end-use materials for both functional prototype evaluation and low-volume production.

Indirect Benefits   There are benefits of RP to marketers and consumers. To the marketer, it presents new capabilities and opportunities. It greatly reduces the time it takes to market a product by having accurate working and non-working models that can be used in focus groups and other market analysis. Other benefits include:

- Reduced risk to market a product due to customer review and feedback
- Marketing products that better suit a customer's needs
- Offering of diverse products

Consumers can buy products of their choice that are more reliable and meet their individual needs. Consumers can collaborate with product developers to make custom products. Since RP reduces waste, manufacturers can pass on the savings to customers.

10.3.3 Lean Production

Since productivity has become such an important aspect of modern manufacturing, there has been a push for systems to increase the productivity of individual companies. Toyota Production Systems, the world's foremost production organization, was the first to develop the principle of lean excellence. When the "lean" principles are applied to manufacturing, they are referred to as lean production. Lean production is lean because it provides a way to do more with less and less—less time, less space, less human effort, less machinery, less materials—while coming closer and closer to providing customers with exactly what they want. These principles have been widely publicized and applied at companies worldwide.

The term *lean production* was coined by MIT researchers to describe the efficiency improvement programs that Toyota Production Systems adopted to survive in the Japanese automobile market after World War II [4, 5].

Lean production can often be described as an adoption of mass production in which workers and work cells are made more flexible and efficient by adopting methods that reduce waste in all phases. Lean production is generally based on four principles [6]:

1. Minimize waste
2. Perfect first-time quality
3. Flexible production line
4. Continuous improvement

The four principles of lean production are discussed next.

Minimize Waste   Although there are four principles of lean production, they are basically derived from the first principle: minimize waste. Taiichi Ohno, the Toyota chief engineer who initiated all the cost-cutting measures to improve the efficiency of the production system, came up with the following list of wastes [7]: (1) production of defective parts, (2) production of more than the number of items needed, (3) unnecessary inventories, (4) unnecessary processing steps, (5) unnecessary movement of people, (6) unnecessary transport of materials, and (7) workers waiting. Procedures were developed at Toyota plants to minimize these forms of waste such as just-in-time (JIT) manufacturing. JIT procedures emphasize the production of no more than the minimum number of parts needed at the next workstation. This reduces unnecessary inventories. Rapid prototyping and rapid manufacturing represent the essence of "lean" such that only the desired product is manufactured when it is needed with almost no wasted material or time. In addition, most RP systems produce their output unattended and thereby release operators to engage in other productive efforts.

Perfect First-Time Quality   Before we talk about this principle of perfect first-time quality, let us talk about the differences between mass production and lean production. Table 10.3 summarizes the differences between the two production systems.

Mass production and lean production define the term *quality* differently. In mass production, quality represents an acceptable level, which means a small fraction of defect is acceptable, even satisfactory. In lean production, perfect quality of zero defects is required. The JIT system practiced in lean production requires a zero defect level in part quality because if the part delivered downstream is defective, the production stops. Since defective parts are not acceptable, lean production emphasizes that the parts be made right the first time. On the other hand, inventory buffers are used in mass production to accommodate quality problems. With RP&M, inventory buffers are unnecessary and prototypes can be easily reviewed well ahead of time to eliminate manufacturing defects.

TABLE 10.3   Comparison of Mass Production and Lean Production

Mass Production	Lean Production
Inventory buffers	Minimum waste
Just-in-case deliveries	Minimum inventory
	Just-in-time deliveries
Acceptable quality levels (AQL)	Perfect first-time quality
Tailor system	Worker teams
Maximum efficiency at the component/	Worker involvement
subassembly level	Flexible production system
"If it ain't broke, don't fix it"	Continuous improvement

Flexible Production Systems   A flexible manufacturing system, or FMS as it is commonly called, is a reprogrammable manufacturing system capable of producing a variety of products automatically. Since Henry Ford first introduced and modernized the transfer line, we have been able to perform a variety of manufacturing tasks automatically. However, incorporating minor changes in the product has been difficult because the system is not flexible enough. We know that in today's competitive market, it is essential to accommodate customer's changes, otherwise the customer will go to someone else who can meet their needs in a timely manner. Since all prototypes are maintained in an electronic CAD file, changes are almost instantaneous and inexpensive.

Mass production systems use transfer lines to produce large volumes of product at low cost. In mass production, the goal is to maximize efficiency, which is achieved using long production runs. Lean production incorporates FMS in its manufacturing philosophy. Toyota needed flexible systems to meet the demand of the smaller car market and to become as efficient as possible. RP&M can exceed the flexibility of FMS if the challenges of more robust materials and high accuracy and volume can be overcome.

Continuous Improvement   In mass production, there is a tendency not to take any action so long as a production system is working. Only when something fails are actions initiated. By contrast, lean production encourages the policy of continuous improvement. Known as *Kaizen* by the Japanese, continuous improvement means constantly searching for and implementing ways to reduce cost, improve quality, and increase productivity [8]. Continuous improvement may involve any or all of these problem areas: cost reduction, quality improvement, cycle time reduction, manufacturing lead time reduction, and work-in-process inventory reduction, as well as improvement in production design to increase performance and customer satisfaction. As we have stated in previous sections, RP&M not only improves the manufacturing process by reducing waste and helping to identify processing problems early, it also improves design efficiency by giving designers easy, cheap, and quick access to physical prototypes. The availability of physical prototypes at the earliest design phases are proved to be of significant value in identifying design flaws and supporting effective use of concurrent engineering and reduced product development cycle times.

10.4   SERVICE BUREAUS

Rapid prototyping is one of the most important aspects of manufacturing. It is a relatively new industry that has greatly impacted the manufacturing process. RP has greatly affected the lead time and cost of manufacturing. Currently, the main drawback with RP is the high cost of purchasing a system. Fortunately, there are alternatives to purchasing a unit. The two main alternatives are service bureaus and consortia. In this section, we shall discuss the features of both the alternatives of buying a RP system [2].

10.4.1   What Is a Service Bureau?

A service bureau is a company that owns and operates one or more rapid prototyping machines and will contract with other companies to use those machines to produce prototypes for them. Service bureaus also offer many other services including concept development, computer-aided design, data translation, prototyping, rapid tooling, casting, reverse engineering, and other engineering and manufacturing services. Of course, the service bureau charges for these services. However, especially in the case of RP parts, the cost will be much less than purchasing a machine. In addition, the customer can easily experiment with various RP technologies before making a purchase decision. Some service bureaus have expertise in many other areas of manufacturing such as CNC machining and injection molding.

Service bureaus play an important role in the rapid prototyping market. The entrepreneurial companies who operate service bureaus are among the leaders in pursuing new applications for the various rapid prototyping technologies. Moreover, service bureaus have introduced many larger companies to rapid prototyping through educational efforts and by building parts for these organizations efficiently and cost effectively.

The first RP service bureaus were formed in the late 1980s. As soon as RP became a common method for creating new models, the need for an outside resource that allowed companies to benefit from RP without purchasing and operating the equipment in-house was introduced. About 400 service providers (bureaus) were working worldwide at the end of 2001, according to CAD/CAM Publishing's *2002 Rapid Prototyping Directory*. This shows a modest growth of 3.4% over the previous year. Service bureaus are currently not doing as well as in the past. They are struggling with lack of incoming projects and cash flow due to the current economic conditions. Many are being forced to downsize and operate very lean. Figure 10.6 shows the location of service providers in regions around the world. The percentage of particular machines being used around the world is shown in Figure 10.7. The number of machines being installed at service providers of the total installed base of RP machines worldwide is shown in Figure 10.8.

Although service providers use a wide variety of RP systems, 3D Systems' stereolithography machines dominate the installed base of all RP systems. Figure 10.7 indicates the percentage of machines used by service bureaus.

The first RP machine, the SLA 1, was used at a service bureau in Illinois. Many other service bureaus began to sprout shortly thereafter, some as divisions of existing corporations, but most as startup companies that bought RP equipment and made this new technology more widely available to the marketplace.

10.4.2   Reasons for Using Service Bureaus

There are many reasons why a company may choose to employ a service bureau to handle its rapid prototyping requirements. The most common reasons [8] are described next.

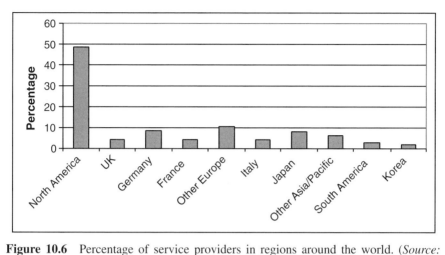

Figure 10.6 Percentage of service providers in regions around the world. (*Source:* Wohlers Report 2002.)

Expertise in Secondary Processes Technology must often be coupled with secondary processes in order to leverage the benefits of rapid prototyping. By serving as advisors and consultants on the best process or combination of processes, service bureaus provide added value to their services. Since service bureaus also deal with a wide variety of end-use applications and service to many industries, more often than not, they possess expertise a single company may not have in-house.

Multiple Technologies Every rapid prototyping technology has its own specific strength, and a company might have a wide variety of needs. For ex-

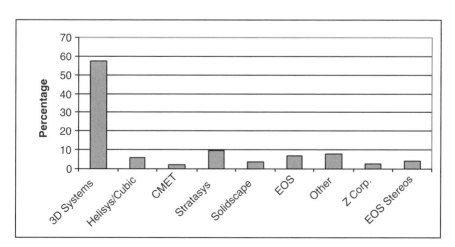

Figure 10.7 Percentage of machines being used around the world. (*Source:* Wohlers Report 2002.)

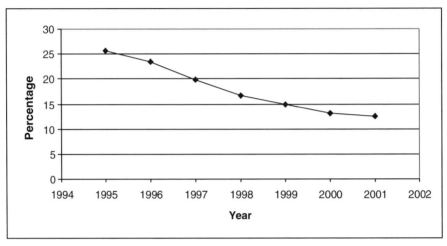

Figure 10.8   Shrinking of market segment. (*Source:* Wohlers Report 2002.)

ample, for a certain application, a ductile part might be needed. However, for another, there may be a need for extreme accuracy and precision. A company can go out to several service bureaus in order to accommodate its needs, instead of having to compromise requirements based on in-house capabilities.

Occasional Prototyping Requirements   The prototyping requirements of some companies may be too small to justify purchasing their own equipment. If a machine is only used half the time, it is probably not worth the investment to the company. Continuing investments in the equipment such as consumables, upgrades, maintenance, and technical expertise are just a few of the costs of owning the machinery. Some companies may need rapid prototyping many times throughout the year; however, the need may come in peaks and valleys. Using a service bureau, or multiple service bureaus, gives a company the flexibility to handle a highly variable workload.

Expense of the Equipment   Many of the companies who need rapid prototyping are small companies that do not have the capital equipment budget to purchase the equipment, upgrades, consumables, or technical expertise. Most rapid prototyping machines are expensive to own and operate. Before going out and purchasing a rapid prototyping machine, it is often beneficial for a company to test the technology as well as the results of various rapid prototyping machines by using a service bureau. By doing so, companies can ensure that the process being evaluated is suited for the application at hand.

Rapidly Changing Technology   As rapid prototyping continues to evolve at an ever-increasing rate, better, faster, less expensive, and more accurate processes are being developed and defined. Many companies feel that at this stage in the development of rapid prototyping, it would be more beneficial to maintain the flexibility to use today's best technology, rather than being locked into what may prove to be an outdated system.

Hazardous Material   Some rapid prototyping processes require the use of hazardous materials, which must be handled and disposed of properly. Many companies have restrictions that relate to the handling of such materials and are not equipped or willing to deal with the costs and potential liability of this issue.

10.4.3   Factors in Choosing a Service Bureau

Appropriate Technology   Every rapid prototyping process has its own set of strengths and weaknesses. Thus, it is important for users to understand the requirements of each systems application and to compare the prototype requirements to the relative capabilities of the available rapid prototyping technologies [8].

Coordinating Secondary Processes   The optimum prototyping solution is usually achieved by combining a rapid prototyping process with a complimentary secondary process such as urethane casting or investment casting. Having a close coordination with secondary processes is vital to the overall success of the project. Service bureaus either handle secondary processes inhouse or through closely managed relationships with outside suppliers such as foundries, spray metal tool shops, platers, and the like. Users can help make sure a project is successful by choosing a service bureau with expertise in the secondary process to be employed, as well as the RP process.

Quality   The quality of parts built by rapid prototyping processes is affected by factors such as layer thickness, build orientation, build parameters, equipment calibration, materials used, finishing techniques, and so forth. Users of rapid prototyping depend on a bureau's expertise in the technologies being offered. In order to ensure that the prototyped parts are of acceptable quality, the user should determine that the service bureau:

- Understands the capabilities and limitations of the rapid prototyping process being proposed.
- Has the ability, experience, and expertise to get the most out of the rapid prototyping process.
- Understands the requirements of the application.
- Is committed to delivering the parts that meet the requirements of the application in a timely manner.

This assurance usually comes from references provided by the service bureau, industry consortia, or user groups of this RP technology.

Competitive Pricing and Turnaround Time   Pricing and turnaround time vary among rapid prototyping service bureaus based on the technologies used, capacity available, and methods for determining costs. Since many rapid prototyping projects are on tight schedules, many rapid prototyping users establish relationships with one or two service bureaus to do the bulk of their work. In these relationships, standard rates and turnaround times can be established early so that time is not wasted when parts are urgently needed.

Observation   Service bureaus represent a magnificent opportunity for businesses and individuals to take advantage of rapid prototyping without incurring the capital and operating costs of owning a rapid prototyping machine. They are also useful to businesses that own their RP equipment for outsourcing overflow work and obtaining special capabilities that cannot be handled by the particular machines they have installed. Finally, they allow prospective purchasers of rapid prototyping machines to evaluate and compare the output of the various machines before committing to a particular technology.

10.5   RAPID PROTOTYPING CONSORTIA

A consortium can be an alternative to a service bureau. There are two main types of consortia: university–industry and industry–industry. The most common type of consortia is university–industry. This type of consortia involves the cooperation of universities and industry to advance RP. The main benefit for the university is the opportunity it provides for the students to be exposed to this technology. For industry, it can be a cheaper alternative than a service bureau. The industry–industry consortium involves only industry; together they look for ways to improve rapid prototyping applications as it relates to their particular industry. Consortia are an important part of the rapid prototyping industry.

10.5.1   Benefits of Consortia Affiliation

There are many benefits associated with consortia affiliation. One benefit is that a consortia provides product development assistance. This is very important to small industrial firms that are looking for ways to improve their products. The next benefit is that they provide seminars and in-plant training in the practical use of RP. When a company first purchases an RP machine, it usually does not know all the capabilities of the machine. Through the consortia, workers can learn to use their RP to the fullest extent. Another benefit is the participation in problem-focused consortium subcommittees. Many companies encounter problems with their RP machines. There may be other companies that are encountering similar problems. In these subcommittees, they can work together to solve those problems. A third benefit is that there are monthly networking sessions and workshops in consortia. In these networking sessions, companies with new RP machines can talk to experienced users to resolve some of their problems. The workshops are useful for companies learning to use RP machines as well as companies looking to improve their current use of their installed RP machines.

10.5.2   University–Industry Consortia

One type of rapid prototyping consortia is called university–industry consortia. This type of consortia is very common. It benefits both the university and the industries involved. The way a university–industry consortia works is that the university has one or several RP machines, and companies pay to have the university produce prototypes for them. Together the university and the industries work together to advance the value of RP to the industrial partners. Another aspect of these consortia is to improve manufacturing methods and manufacturing processes. One important aspect of university involvement is research. Through university research, companies can find new ways to improve their methods and processes without the expense and limitations of in-house research. By improving their methods and processes, companies can save a lot of money, thereby offsetting the cost of the consortium membership.

One example of a university–industry consortium is called the Rapid Prototyping Center (RPC), located at the Milwaukee School of Engineering. The mission statement for the center is as follows:

- To research and develop new manufacturing methods and processes related to rapid prototyping
- To reduce financial risk for companies by implementing rapid prototyping as a means to manufacture parts and prototypes
- To assist students in their educational development who will be ready to implement new technological practices upon entry into industry
- To inform and educate industry, as well as the general population, of upcoming rapid prototyping technologies and processes

For the university, the most important aspect is the experience it provides its students. The students who are involved in the RPC gain invaluable experience that they can use upon graduation. For industry, the importance is two-fold. The first important aspect is helping companies that are implementing RP for the first time. Since RP implementation is so expensive, companies can learn how to use the RP system to its full capabilities in a more efficient manner. The second important aspect involves learning about new ways to improve rapid prototyping within their companies.

Currently, Loyola Marymount University (LMU) functions as a consortia with some local divisions of the major aerospace companies. Dr. Rafiq Noorani is in charge of the rapid prototyping at LMU. The RP activities conducted at LMU consist of research and consulting. During the summer, research is conducted to improve rapid prototyping. This research is conducted under a grant from the National Science Foundation. The consulting is conducted throughout the year. Since the machine is idle when no research is being conducted, companies may pay to have parts manufactured at LMU. LMU provides companies an alternative to service bureaus. The price of a service bureau is generally much higher than that of a consortia, and the services provided by LMU to local partners is no exception.

10.5.3   Industry–Industry Consortia

Another type of consortium is called industry–industry. This type of consortia involves companies from similar industries. One of the goals of this type of consortia is to reduce engineering development time. In industry, lead time is extremely important. By reducing lead time, industries can enter the marketplace a lot faster. The second goal of this type of consortia is to lower cost. All industries are looking to lower the cost of manufacturing. The reduction in lead time is associated with lowering cost by reducing engineering development time. Another goal is to implement new advances in RP. A lot of research is currently being done in the rapid prototyping industry. The advances that have been discovered can be more readily implemented to improve the technology and overall profitability of the industry partners.

10.6   PRESENT AND FUTURE TRENDS

Modern rapid prototyping techniques and conventional prototyping techniques such as CNC machining and casting are the key technologies for improving the product development process and reducing lead time to market. Design and manufacturing industries are using rapid prototyping technology for product development in ever-increasing numbers.

More than one million prototypes are now made each year, and, with more than a million engineering workstations installed, the trend is expected to continue. It is also expected that the worldwide sales of rapid prototyping businesses will exceed $1 billion 2005.

A significant portion of the early rapid prototyping business was primarily concerned with providing physical models of CAD to help designers and manufacturers in the development of the product. However, newly emerging RP products are being targeted toward better defined niche applications, for example, casting, rapid tooling, and medical applications.

The main competition to rapid prototyping systems is NC (numerical control) machining. NC machining is normally used in small and medium-sized model making shops. Compared to current RP systems, NC machining produces parts with far greater accuracy and has a much broader range of materials. Some of the challenges of current rapid prototyping systems are their inability to:

1. Produce parts with the accuracy and surface finish required for engineering models
2. Prototype with a wide variety of engineering materials
3. Directly produce high-quality metal parts for production tooling applications
4. Prototype parts as fast as conventional machining

Rapid prototyping made from various processes and materials has shrinkage factors associated with each prototype when the materials undergo a phase change from liquid to solid [9]. It is essential to accurately predict the shrink compensation factor so that the part is produced as accurately as possible from the design database [10]. The accuracy of processes such as stereolithography and selective laser sintering were not very good when they were first introduced. However, the accuracy of all RP systems is being improved because of the demands to create precision molds that have unusual curves and complex shapes. Accuracy of parts may vary from ± 25 to 75 μm (0.001 to 0.003 in.) for simple parts to as high as ± 125 to 375 μm (0.005 to 0.015 in.) for complex parts. The Solidscape (previously known as SPI) RP systems have accuracy generally 10 times higher than FDM and other RP systems.

Efforts are also being made to increase the strength of RP parts since they are usually weaker than machined parts. The parts that are made by FDM machines using ABS plastic are very strong compared to parts made from thermoset materials. Recent work of Cheung and Ogale [11] has shown an increase in strength of photopolymers by fiber reinforcement. DuPont researchers have shown similar increase in strength of thermoset resins used in stereolithography machines. Also, a new process called laser engineering net shaping (LENS) has been developed at Sandia National Laboratory that permits direct fabrication of high-strength metal parts.

Rapid prototype, as the name implies, should involve prototyping parts rapidly. Some RP systems are not rapid enough to compete with conventional machining systems. For example, Solidscape ModelMaker-6B is probably the slowest RP system available on the market. On the other hand, the accuracy of the ModelMaker-6B is probably the highest of all of the available RP systems.

Rapid prototyping systems employ additive processes to build mostly monomaterial prototypes [12]. Efforts are being made, however, to combine the benefits of both additive and subtractive processes. Rapid tooling has been the main application area so far. The *Fraunhofer IPT* has developed an experimental system, called "laser generated RP," that uses laser welding to melt metal powder as it drops from a coaxial laser/powder distribution cone.

The combined process of addition and subtraction described above does not include support structures. Efforts are underway at Carnegie Melon and Stanford Universities to develop a process called shape deposition manufacturing (SDM), which will include support structures. In this system, the CAD model is first sliced into 3D layered structures that are then deposited as near net shapes and then machined to net shape before additional material is deposited. The sequence for depositing and shaping the build and support materials depends on the local geometry. The idea here is to decompose shapes into layer segments such that undercut features are formed by previously shaped segments.

Based on the above discussions, it can be safely said that rapid prototyping and its associated technologies are having a profound impact on the way companies produce models and prototype parts. The technology is also impacting the lives of people who have decided to use, manage, teach, or develop various aspects of rapid prototyping or rapid tooling. Product designers, manufacturers, educators, researchers, and countless others are working hard to understand this technology and increase its capabilities in the years ahead.

10.7   SUMMARY

In this chapter, we presented mainly the guidelines for implementing RP and managing issues relating to this new technology. RP is still an expensive investment. Purchasing decisions should be based on sound criteria, such as cost trade-off, cycle time, accuracy and surface finish of prototypes, and size and strength of the part. When a purchasing decision is taken, considerations for facility planning and environmental factors should be taken into account.

This chapter then presents the operating issues that include equipment cost, building cost, training, and maintenance cost. Since RP is a technology in itself, it draws on other technologies, such as CAD/CAM, CNC machining, and so forth. Once RP is integrated into the company, many direct and indirect benefits from this technology can be obtained. Information on how RP contributes to lean production are also presented in this chapter.

Purchasing an RP system may not be feasible for every company, especially a small company. Service bureaus and consortia are organizations that provide the RP needs of companies. This chapter also describes what they are, how they operate, and the criteria for selecting these organizations that provide easy and efficient access to RP technology. The chapter ends with some comments on the present and future trends of RP.

PROBLEMS

10.1. Discuss the options a company has for using rapid prototyping system as part of their product development system.

10.2. Mention the criteria to be evaluated by a company when selecting a particular type of rapid prototyping process.

10.3. What are the criteria to be used in evaluating a rapid prototyping system.

10.4. Mention the criteria to be selected in evaluating a service bureau? Is there any hidden costs? If so, what are they?

10.5. What are the new trends in RP system?

10.6. RIN Corporation plans to purchase a rapid prototyping system at a price of $120,000. The first-year operation and maintenance cost is $30,000. The Company is eligible for $20,000 tax credit from the federal government under its technology investment program. The RP system replaces 5 design and analysis workers. The hourly rate for the worker is $20 including fringe benefits. Determine the payback period for the company.

REFERENCES

1. K. Otto and K. Wood, *Product Design: Techniques in Reverse Engineering and New Product Development,* Prentice-Hall, Upper Saddle River, NJ, 2001.
2. C. C. Kai and K. F. Leong, *Rapid Prototyping: Principles and Applications in Manufacturing,* Wiley, Singapore, 1997.
3. M. P. Groover, *Automation, Production Systems and Computer-Integrated Manufacturing,* Prentice-Hall, Upper Saddle River, NJ, 2001.
4. K. Womack, D. Jones, and D. Roos, *The Machine That Changed the World,* MIT Press, Boston, MA, 1990.
5. K. Womack and D. Jones, *Lean Thinking,* Simon and Schuster, New York, 1996.
6. D. Roos, Agile/Lean: A Common Strategy for Success, *Agility Forum,* 1995.
7. T. Ohno, *The Toyota Production System: Beyond Large Scale Production,* Productivity Press, Portland, OR, 1988.
8. P. F. Jacobs, *Stereolithography and Other RP&M Technologies from Rapid Prototype to Rapid Tooling,* ASME Press, New York, 1996.
9. B. Fritz and R. Noorani, Form Sheet Metal with RP Tooling, *Advanced Materials & Processes,* Materials Park, Ohio, **155**(4), 37–42, 1999.
10. Q. Dao, J. Frimodig, H. Le, X. Li, S. Putnam, K. Golda, J. Foyos, R. Noorani, and B. Fritz, Calculation of Shrinkage Compensation Factors for Rapid Prototyping, *Computer Applications in Engineering Technology,* **7**(3), 1999.
11. T. Cheung and A. Ogale, Processing of Multi-layer Fiber Reinforced Composites by 3D Photolithography, in Proceedings of the 1998 NSF Grantees Design and Manufacturing Conference, Monterey, Mexico, 1998, pp. 557–559.
12. J. Bohn, File Format Requirements for the Rapid Prototyping Technologies of Tomorrow, International Conference on Manufacturing Automation Proceedings, Hong Kong, 1997.

11

RESEARCH AND DEVELOPMENT

The success of rapid prototyping depends upon the continued research and development in the fields. These challenges include improving the rapid prototyping processes as well as improving part accuracy, build speed, and material properties. This chapter presents examples of research and development efforts of the author and his students at LMU by way of illustration of the type of research that is both useful to the industry and instructive for the student. There are many excellent research projects being done in both academia and industry, and it is left to the individual instructor to choose the most appropriate and timely examples available to illustrate pertinent concepts at the time the course is presented.

11.1 IMPROVEMENT OF FDM PROCESS USING DESIGN OF EXPERIMENT

11.1.1 Abstract

The purpose of this research was to improve the fused deposition modeling process by examining the tensile strength of samples fabricated in a Stratasys FDM 1650 machine utilizing the methods of design of experiments. A 2-level, 4-factor, full factorial design was deemed the most appropriate for this experiment. The 4 factors selected were temperature, air gap, slice thickness, and raster orientation. There were 16 trials following the full factorial matrix. Three samples of each trial were made for a total of 48. Each sample was stressed to the point of fracture using a tensile test. An average was taken for the yield and ultimate stress and used in the response tables. From the response table, effects graphs, and normal probability graphs, a regression equation was found in order to determine the level at which each factor should be set to optimize the FDM machine to produce high tensile strength samples. It was found that small air gap, small slice thickness, low raster orientation 0°/45°, and the interaction between temperature and slice thickness were the factors that yielded the strongest samples.

11.1.2 Introduction

Fused deposition modeling (FDM) is a process that is used for fabricating solid prototypes from a computer-aided design (CAD) data file [1]. The process fabricates 3D parts from a buildup of 2D layers. In this process, an acrylonitrile–butadiene–styrene (ABS) thermoplastic polymer is extruded through a heated nozzle to deposit the layers. In a previous study, we showed that the orientation of the layers created anisotropic behavior in the tensile strength [2]. Other investigators have also experienced similar results [3].

Previous work has used design of experiments (DOE) as a method to maximize strengths of the FDM-processed specimens of silicon nitride [4] and ABS polymer [3]. However, many of the factors used for influencing the strength were entirely different in these studies. In addition, these statistically designed experiments on FDM processing have not been physically interpreted in terms of the material properties and microstructure.

Even though there are many advantages, material property is a challenge that must be faced. In the early years of RP, materials had very low yield strength. Even though materials have gotten sturdier, the characteristics of the material used in RP generally exhibit weak yield strength. The goal of engineers is to be able to test prototypes in wind tunnels at high speeds of velocity. If this is accomplished, this will allow different opportunities in technological advances.

The purpose of this experiment is to increase the tensile strength of rapid prototyped specimens that are manufactured in an FDM machine. The yield and ultimate stress will be analyzed utilizing the methods of DOE. The DOE technique is an approach to statistically optimize the response value by selecting two-level parameters for various factors. It is a technique based on systematically varying the levels of independent variables. This method allows several variables along with their interactions to be studied in a systematic manner.

Careful analysis of the data can provide invaluable information about how the input variables affect the responses (quality characteristics). This can result in significant improvements to products and processes, including shorter development cycles, more robust products, variation decrease, improved customer satisfaction, and cost reductions. An experiment designed and conducted properly eliminates all possible causes of variation, except the one that we are studying. Therefore, the knowledge and application of DOE is critical for manufacturers, developers, and researchers in industry.

This project will involve selecting a high and low level of four parameters. These factors are temperature, air gap, slice thickness, and raster orientation. The factors will be combined in all possible combinations and tensile specimens will be made and tested. The yield and ultimate stress will be analyzed using a classical DOE method. The data gathered will allow the effects of each factor and their interactions to be evaluated to determine which produce the strongest specimen.

The purpose of this chapter is to use the quality engineering tools to design, analyze, and physically interpret our selection of the FDM processing factors and their levels.

11.1.3   Design of Experiments

Many different factors (parameters) go into the fabrication of one solid model. When several factors affect the response (y), then the best strategy of experimentation is to vary all the factor combinations together. The data is statistically analyzed to determine the affects of the factors and their interactions on the response. The method known as design of experiments (DOE) is utilized in this project and intended to help improve the quality of a design and process by using statistical analysis. Improving an experiment is to optimize the average response value and minimize the affects of variability on process or product performance. An experiment can be defined as a series of trials or tests that produce quantifiable outcomes.

Design of experiments has provided documented substantial savings to thousands of companies by solving difficult quality problems, reducing product and process variation, and optimizing product/process performance and consistency. DOE is a very powerful analytical method that can be taught to industry professionals at a very practical level, providing a cost-effective and organized approach to conducting industrial experiments. A major benefit of DOE is that multiple product design and/or process variables can be studied at the same time with these efficient designs, instead of a hit-and-miss approach, providing very reproducible results. Due to the statistical balance of designs, thousands of potential combinations of numerous variables can be evaluated for the best overall combination, with a minimal amount of trials.

Taguchi experimental designs (also called robust design in the United States) are based on orthogonal arrays and were made popular by the Japanese engineer Genichi Taguchi [5]. They are identified with a notation such as L_8, which indicates an array with 8 trials as displayed in Figure 11.1. It uses 1 to indicate a low level and 2 to indicate a high level. Taguchi ignores interactions between factors. These methods integrate statistical design of experiments into a powerful engineering process. The simplicity of implementation has attracted Japanese manufacturers to improve their product and their processes.

Taguchi L$_8$ Design						
1	**2**	**3**	**4**	**5**	**6**	**7**
1	1	1	1	1	1	1
1	1	1	2	2	2	2
1	2	2	1	1	2	2
1	2	2	2	2	1	1
2	1	2	1	2	1	2
2	1	2	2	1	2	1
2	2	1	1	2	2	1
2	2	1	2	1	1	2

(Trial column: 1, 2, 3, 4, 5, 6, 7, 8)

Figure 11.1   Taguchi design matrix.

Classical experimental designs are similar to the Taguchi method in that they are also based on orthogonal arrays but instead are identified with a superscript to indicate the number of variables. A 2^3 classical experiment such as the one used in this project has 4 factors of 2 levels each, producing a total of 18 trials or experiments. The classical method uses a -1 to indicate the low level and a $+1$ to indicate the high level. As seen in the design matrix of Figure 11.2, interactions between factors are taken into account so further analysis can be done.

For this experiment the classical method was used in order to analyze the affects of the interactions between the various factors. There are two common ways of conducting a design of experiments. The first is using a full factorial design where each factor A, B, and C is analyzed along with all possible interaction combinations. In order to conserve time, a less accurate method known as a fractional factorial design is used. However, because time was not an issue, a full factorial is used in this project so all possible interactions could be studied. The factors that effect the information in the experiment are supported by the following fact: The standard errors of most estimators are

Classical 2^3 Design						
A	**B**	**C**	**AB**	**AC**	**BC**	**ABC**
-1	-1	-1	1	1	1	-1
1	-1	-1	-1	-1	1	1
-1	1	-1	-1	1	-1	1
1	1	-1	1	-1	-1	-1
-1	-1	1	1	-1	-1	1
1	-1	1	-1	1	-1	-1
-1	1	1	-1	-1	1	-1
1	1	1	1	1	1	1

(Trial column: 1, 2, 3, 4, 5, 6, 7, 8)

Figure 11.2   Classical design matrix.

proportional to σ (a measure of data variation) and inversely proportional to the sample size.

11.1.4 Selection of Build Parameters

The process control parameters that were most likely to have an impact on the tensile strength of the FDM tensile specimens were identified. These parameters or variables selected are chosen based on intuitive knowledge about the FDM process and the ABS plastic. The four quantitative parameters chosen to study are temperature (A), air gap (B), slice thickness (C), and raster orientation (D). Controlling the information in an experiment is very important. The classical DOE method will help to interpret the data that is gathered from the tensile specimen. A design matrix is shown in Figure 11.3 to organize all possible combinations of the four-factor (A, B, C, D), two-level, three-level, and four-level interactions.

These four process control parameters were identified as being the most likely to affect the properties of an FDM model. Discussed below is a brief description of each parameter.

> *Factor A: Model Temperature* The temperature of the heating element from which the ABS plastic is extruded through. This controls how molten the material is as it is deposited. Figure 11.4 shows the process by which the ABS plastic is extruded through the nozzle of the FDM machine.
>
> *Factor B: Air Gap* This is the space between the beads of the FDM material. The default is zero, placing the beads alongside each other. A positive air gap leaves a space or gap between the beads. A loosely

Trial	A	B	C	D	AB	AC	AD	BC	BD	CD	ABC	ABD	ACD	BCD	ABCD
1	-1	-1	-1	-1	1	1	1	1	1	1	-1	-1	-1	-1	1
2	-1	-1	-1	1	1	1	-1	1	-1	-1	-1	1	1	1	-1
3	-1	-1	1	-1	1	-1	1	-1	1	-1	1	-1	1	1	-1
4	-1	-1	1	1	1	-1	-1	-1	-1	1	1	1	-1	-1	1
5	-1	1	-1	-1	-1	1	1	-1	-1	1	1	1	-1	1	-1
6	-1	1	-1	1	-1	1	-1	-1	1	-1	1	-1	1	-1	1
7	-1	1	1	-1	-1	-1	1	1	-1	-1	-1	1	1	-1	1
8	-1	1	1	1	-1	-1	-1	1	1	1	-1	-1	-1	1	-1
9	1	-1	-1	-1	-1	-1	-1	1	1	1	1	1	1	-1	-1
10	1	-1	-1	1	-1	-1	1	1	-1	-1	1	-1	-1	1	1
11	1	-1	1	-1	-1	1	-1	-1	1	-1	-1	1	-1	1	1
12	1	-1	1	1	-1	1	1	-1	-1	1	-1	-1	1	-1	-1
13	1	1	-1	-1	1	-1	-1	-1	-1	1	-1	-1	1	1	1
14	1	1	-1	1	1	-1	1	-1	1	-1	-1	1	-1	-1	-1
15	1	1	1	-1	1	1	-1	1	-1	-1	1	-1	-1	-1	-1
16	1	1	1	1	1	1	1	1	1	1	1	1	1	1	1

Figure 11.3 Four-factor–two-level orthogonal matrix.

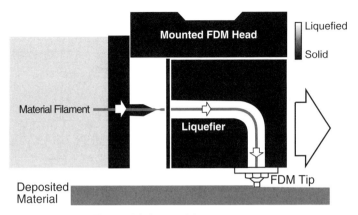

Figure 11.4   Model temperature.

packed structure is created with poor strength. A negative air gap results in beads that partially occupy the same space forming a much stronger arrangement. (Fig. 11.5).

Factor C: Slice Thickness   This is the thickness of the layer laid down. The larger the slice the more rapid the model is built. However, more slices yield a stronger more accurate part. Figure 11.6 shows three layers of the specimen.

Factor D: Raster Orientation   The direction of the beads of material (roads) relative to the loading of the part. Figure 11.7 displays two different raster orientations tested.

Table 11.1 shows the high and low level of each parameter that was varied.

11.1.5   Experimental Method

There were six major steps ranging from designing and constructing the tensile specimens to testing the specimens and analyzing the data that was gathered. These steps more specifically included selecting parameters and levels, a CAD process, setup and build, postprocessing, tensile testing, and data analysis. These steps are summarized below:

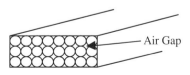

Figure 11.5   Air gap.

Figure 11.6   Slice thickness.

1. Selecting parameters and levels:
   a. Quickslice 6.4 and the FDM 1650 machine were studied in order to find which parameters affect the tensile strength of the samples the most.
   b. Four parameters were selected, choosing a high and low for each. These parameters included: temperature, slice thickness, air gap and raster orientation.
2. CAD process:
   a. A detailed CAD drawing of a tensile specimen was designed in AutoCAD R13.
   b. The CAD drawing was converted into a .stl file for slicing with the Quikslice 6.4 software. It was then converted into a .sml file that could be recognized by the FDM 1650 machine for the building of the part.

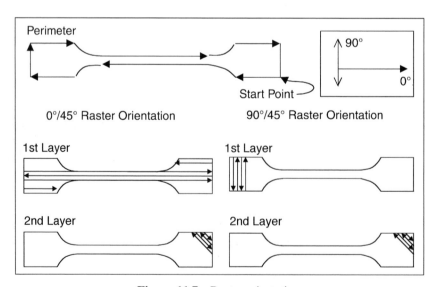

Figure 11.7   Raster orientation.

TABLE 11.1 High and Low Levels Assigned to Each Parameter

Parameter (Factor)	Description	Low Level (−1)	High Level (+1)
A	Temperature	268°C	277°C
B	Air gap	−0.0254 mm	0
C	Slice thickness	0.254 mm	0.356 mm
D	Raster orientation	0°/45°	90°/45°

3. Setup and build: The FDM machine was set for each run based on the four parameters designated to that particular trial number, for a total of 16 trials. The high and low for each parameter are explained below:

   a. Temperature: The FDM machine operated at an overall temperature (envelope temperature) of 68°C. The support material head operated at a temperature of 265°C. The model extrusion head was operated at 268°C for the low and 277°C for the high.

   b. Air gap: The air gap is the space between the beads of the ABS material. The default setting of zero was selected for the high, meaning that the beads just touch. A low value of −0.0254 mm (−0.001 in.) was selected, meaning that two beads partially occupy the same space.

   c. Slice thickness: The slice thicknesses selected were 0.254 mm (0.010 in.) for the low, and 0.356 mm (0.0140 in.) for the high.

   d. Raster orientation: The two layer orientations selected were 0°/45° for the low and 90°/45° for the high, relative to the x axis. The two numbers indicate that layers alternate in their direction as the specimen is constructed. Three samples of each of the 16 trials were made. The trials are shown in Table 11.2. Each trial was setup accordingly and sent to the FDM machine via an SML file.

4. Postprocessing:

   a. The specimens were removed from the FDM machine. Support material was removed using small hand tools such as a razor blade and sanding block.

5. Tensile testing:

   a. Each sample's strength was tested using the Instron Universal Testing Machine 4505.

   b. A stress–strain plot was constructed using Instron software for each sample.

   c. An average was taken in order to find the maximum strength for each run.

TABLE 11.2   Full Factorial Levels for Each Factor in 2^4 Experimental Trials

Random Order	Trial	Temperature A (°C)	Air gap B (in.)	Slice Thickness C (in.)	Raster Orientation D (deg/deg)
14	1	268	−0.001	0.010	45/0
7	2	268	−0.001	0.010	45/90
1	3	268	−0.001	0.014	45/0
12	4	268	−0.001	0.014	45/90
6	5	268	0.000	0.010	45/0
4	6	268	0.000	0.010	45/90
2	7	268	0.000	0.014	45/0
9	8	268	0.000	0.014	45/90
5	9	277	−0.001	0.010	45/0
3	10	277	−0.001	0.010	45/90
11	11	277	−0.001	0.014	45/0
8	12	277	−0.001	0.014	45/90
10	13	277	0.000	0.010	45/0
15	14	277	0.000	0.010	45/90
16	15	277	0.000	0.014	45/0
13	16	277	0.000	0.014	45/90

6. Data analysis:
   a. A response table was constructed, using Design Ease. From this response table, a normal probability graph of effects, a graphical representation of the factor effects, and a graphical representation of interaction effects were plotted.

11.1.6   Results and Analysis of DOE

The first step of the project was to determine the processing factors. Previous experiments have shown that different air gaps affect the strength of ABS polymer. Interactions were suspected between air gap and temperature/layer thickness/raster orientation. Therefore, these four factors were chosen with two levels, a low (−1) and high (1) for each.

An L_{16} (2^4) orthogonal design matrix was utilized using the classical method. To avoid bias, the runs were randomized as in Table 11.2. In order to obtain more accurate data, each trial had three replications. The average was calculated for the strength data, $\overline{Y}_{yield}$ and $\overline{Y}_{ultimate}$. Table 11.3 displays the results.

In order to distinguish which factors and interactions had significant effect on maximum strength, average strength ($\overline{Y}$), effect (the difference between the high and low level of each factor and interaction), standard deviation Si of each run, variance Si^2 of each run, and population standard deviation (σ)

TABLE 11.3 Partial Response Matrix with Average Tensile Strength Data Shown for Each Trial

Trial	A	B	C	D	Y (Yield) (MPa)	Y (Ultimate) (MPa)
1	−1	−1	−1	−1	15.70	16.31
2	−1	−1	−1	1	12.62	13.64
3	−1	−1	1	−1	11.99	12.86
4	−1	−1	1	1	9.06	10.36
5	−1	1	−1	−1	12.97	13.38
6	−1	1	−1	1	12.98	13.81
7	−1	1	1	−1	11.91	12.83
8	−1	1	1	1	7.67	8.77
9	1	−1	−1	−1	17.56	17.77
10	1	−1	−1	1	18.70	19.00
11	1	−1	1	−1	10.31	11.39
12	1	−1	1	1	9.17	10.35
13	1	1	−1	−1	13.70	13.90
14	1	1	−1	1	12.53	13.64
15	1	1	1	−1	10.60	11.36
16	1	1	1	1	6.67	8.83

were calculated. A 95% ($\overline{Y} \pm 2\sigma$) confidence interval for standard error limits was used to verify statistic significance of each factor and interaction [5]. Any factor or interaction that lies outside the interval is significant; otherwise it was ignored. For yield strength, the interval was 12.13 ± 0.73; for ultimate strength, the interval was 13.01 ± 0.73. It is evident from Figure 11.8 that $B_{−1}$, $C_{−1}$, $D_{−1}$, and $(AC)_{−1}$ had significant effect on the yield strength. Figure 11.9 shows that $B_{−1}$ and $C_{−1}$ had significant effect on the ultimate strength.

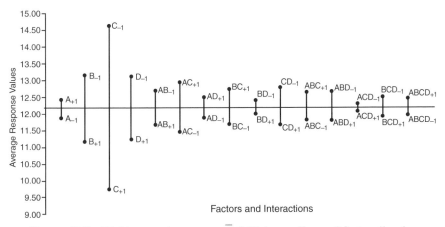

Figure 11.8 Yield strength response $\overline{Y}$ (MPa) vs. effects of factors/levels.

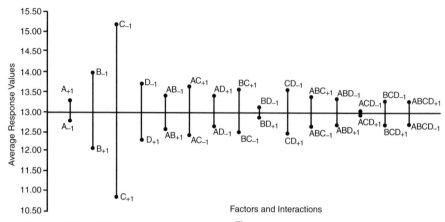

Figure 11.9 Ultimate strength response $\overline{Y}$ (MPa) vs. effects of factors/levels.

The goal of the experiments was to maximize the tensile strength for the various factors and levels. To decide which exact AC interaction gave the biggest yield strength, average values of $A_{-1}C_{-1}$, A_1C_1, A_1C_-, and $A_{-1}C_1$ were calculated and plotted in Figure 11.10. It was concluded from the plot that $A_{-1}C_1$ (or A_2C_1 by Taguchi notation) was the contributing interaction.

A regression equation (Equation 11.1) is another way to find out the level of the factors and interactions that give the maximum strength:

$$\overline{Y}_{max} = \beta_0 + \Sigma_i\beta_i x_i + \Sigma_i\Sigma_j\beta_{ij}x_i{}^*x_j \qquad (11.1)$$

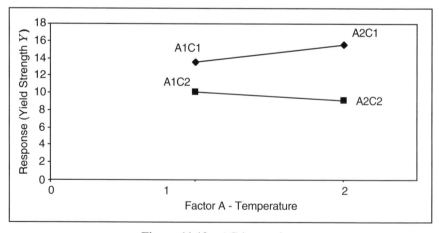

Figure 11.10 AC interaction.

where β_0 = coefficient is the mean value of $\overline{Y}$ (the intercept)
$\qquad \beta_i$ = slope of significant factors
$\qquad \beta_{ij}$ = slope of significant interactions
$\qquad x_i$ = ± 1; represents the low/high level of each significant factor
$\qquad x_{ij}$ = ± 1; represents the low/high level of each significant interaction

Since B, C, D, and AC were known to be the only significant factors and interaction, their effects and the average value of $\overline{Y}$ in the response tables of yield strength and ultimate strength were used in the equation. The highest result was obtained by using B_{-1}, C_{-1}, D_{-1}, and A_1C_{-1}, which just verified the result in Figure 11.8 yield strength response $\overline{Y}$ (MPa) vs. effects of factors/levels:

$$
\begin{aligned}
\text{Yield } \overline{Y}_{max} &= \beta_0 + \Sigma_i\beta_i x_i + \Sigma_i\Sigma_j\beta_{ij}x_i{}^*x_j \\
&= \beta_0 + \beta_B x_B + \beta_C x_C + \beta_D x_D + \beta_{AC}x_{A1}{}^*x_{c-1} \\
&= 12.13 + (-2.01/2)(-1) + (-4.92/2)(-1) + (-1.92/2)(-1) \\
&\quad + (-1.51/2)(1)(-1) \\
&= 17.31 \text{ MPa}
\end{aligned}
$$

With the same equation,

$$
\begin{aligned}
\text{Ultimate } \overline{Y}_{max} &= 13.01 + (-1.90/2)(-1) + (-4.34/2)(-1) \\
&\quad + (-1.42/2)(-1) + (-1.26/2)(1)(-1) \\
&= 17.47 \text{ MPa}
\end{aligned}
$$

11.1.7 Factor Analysis and Physical Interpretation

Microscopic Analysis   In microscopic analysis of fracture surfaces of the samples, we used an optical microscope with magnification ranging from 6.5× to 45×. Pairwise comparisons of samples were conducted in such a way that only one factor would vary in a pair. Using six major characteristics of fracture surfaces, the physical affects of individual variables were determined, for example, bead shape and degree of parallel fiber bonding. Pictures were also taken of the samples' surfaces to help discuss appropriate effects of factors.

Factor Analysis

Factor A: Model Temperature   Higher temperature allowed longer time for viscous flow of ABS molecules. It also contributed to larger kinetic energy of molecules inside the newly laid fibers, which allowed for greater degree of polymer bonding between the fibers. Although by itself the temperature did not appear to be a strong factor (see Figs. 11.8 and 11.9: effects and response graphs), its interaction with slice thickness had a significant affect on the yield strength.

Factor B: Air Gap   When the air gap was set to negative values, the nozzle came closer to the surface of the preceding layer than with the default setting (zero air gap). This resulted in a changed cross-sectional shape of fibers making them more elliptical or in some cases even close to rectangular shape. This increased bonding between parallel fibers of each layer and allowed for a more tightly packed structure by reducing the air spaces inside the composite. Hence, the strength is expected to increase for negative air gaps.

Factor C: Slice Thickness   By decreasing the sizes of individual fibers and thus increasing their number, we were able to produce a composite structure of the same size that was more tightly packed. From samples such as those in Figures 11.11 and Figure 11.12, it was concluded that smaller slice thickness decreased the air space between parallel fibers in individual layers, strengthening the overall composite structure.

Factor D: Raster Orientation   When the two raster orientations are compared, the difference between them was in the 0° and 90° fibers since the 45° layers acted similarly in both cases. In yield and ultimate strength analysis, it is important to notice that difference in response resulted from different orientations of load application. Thus from Figures 11.8 and 11.9, we determined that the composite can withstand a larger load when it is applied along the fibers rather then perpendicular to them. Also it appeared that maximum strain before fracture is much higher for 0°/45° samples.

AC Interaction   From Figure 11.10, it was concluded that temperature affected small slice thickness (C_{-1}) and large slice thickness (C_{+1}) sam-

Figure 11.11   Trial 1.

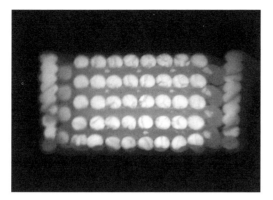

Figure 11.12   Trial 3.

ples differently. While higher temperature increased strength of smaller slice samples, it decreased that of larger slice thickness ones. Our explanation resulted from careful microscopic investigation of appropriate samples. This is shown in Figures 11.11–11.14.

It was observed that in smaller slice samples higher temperatures resulted in closer contact between parallel fibers assisting them in creating a single sheet of material instead of an array of fibers strengthening the weak directions in between the fibers. However, a more liquefied (high temperature) material in large slice thickness samples enabled the molecules to flow into the pores between parallel fibers of the preceding layer. Thus, a resulting fiber came out thinner directly above the previous fibers and thicker in between. When samples were put in tension, these fibers experienced stress concentrations in thin regions and fractured at lower loads than the uniform thickness (lower temperature fibers).

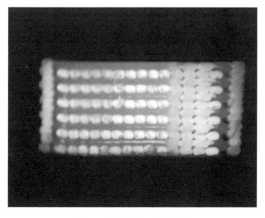

Figure 11.13   Trial 9.

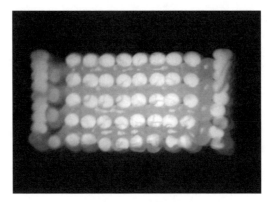

Figure 11.14 Trial 11.

11.1.8 Conclusions and Recommendations

From the design of experiment for FDM–ABS P400, the following conclusions were made:

1. Yield strength is affected by air gap, slice thickness, raster orientation, and the interaction between mold temperature and slice thickness.
2. Ultimate strength is affected by air gap and slice thickness.
3. The low level of every single factor maximizes the strength, so does the interaction between the high-level temperature and the low-level slice thickness.
4. Fracture initiation from large pores located perpendicular to the tensile axis affected the yield strength.
5. Fracture propagation was strongly related to the raster orientation.

For future study, we suggest proportioning the width of road filling with the width of the bead width for 0° orientation. In other words, the filling's width is a multiple of the bead width. This would minimize the size of the pores along the roads and help to achieve more accurate data.

11.2 IMPROVEMENT OF PART ACCURACY

11.2.1 Abstract

Over the past few years, improvements in equipment, materials, and processes have enabled significant improvements in the accuracy of fused deposition modeling (FDM) technology. This project will investigate the present in-plane accuracy of a particular FDM machine using the benchmark "user part" developed by the North American Stereolithography User Group (NASUG) and show the effect of optimal shrinkage compensation factors (SCF) on the accuracy of the prototyped parts.

The benchmark parts were built on the FDM-1650 prototyping machine, and a total of 46 measurements were taken in the X and Y planes using a Brown & Sharpe coordinate measuring machine (CMM). The data was then analyzed for accuracy using standard formulas and statistics, such as mean error, standard deviation, residual error, root mean square (rms) error, and so forth. The optimal SCF for the FDM-1650 machine was found to be 1.007, or 0.7%.

11.2.2 Introduction

Rapid prototyping is the fabrication of a physical, three-dimensional part of arbitrary shape, directly from a numerical description (typically a CAD model) by a quick, highly automated, and flexible process. RP technology has made many contributions to the manufacturing industry mainly by reducing the time to produce prototype parts and improving the ability to visualize part geometry. The physical prototype allows for earlier detection and reduction of design errors and the capability to compute mass properties of components and assemblies [6]. Also, it is very advantageous to present a design in client presentations, consumer evaluations, bid proposals, and regulation certification. This concept was first introduced in 1988, and since then the growth of rapid prototyping has been exponential every year.

Rapid prototyping also has some challenges that must be improved upon, and one of the main challenges is part accuracy. This is the main concern of countries such as Japan, which are not willing to incorporate RP to their manufacturing process [7–9]. The parts produced tend to warp and/or shrink from its given dimensions, forcing the user to run several trials of a part to reach its ideal dimensions or settle for a slightly inaccurate part. Another concern is the build time; it can often take several days to prototype certain parts depending on the size.

When rapid prototyping was first introduced, the materials used to produce the parts had low yield strength. Through the advancement in material science, the photopolymers and thermoplastics used now have much higher yield strength. The strength of RP materials is sufficient for some applications but does not always satisfy the strength requirements for wind tunnel testing. The use of stronger materials in the future such as metal-based powders will enable RP to produce parts to satisfy these requirements.

The purpose of this engineering research project is to make a contribution toward the accuracy improvement of rapid prototyping machines. This can be accomplished by finding the most optimal shrinkage compensation factor. The SCF determines the amount an object should be enlarged so that it can approach the desired dimensions of the user during the shrinking process.

11.2.3   Theory and Analysis

In this engineering project, the rapid prototyping technology used was an FDM-1650 machine built by Stratasys. The machine uses a 3D CAD drawing that is converted to an .STL file, and then it is sliced into a series of closely spaced horizontal planes [10, 11]. The file is then sent to the machine as a build file and production of the part begins. The machine has a build area of $10 \times 10 \times 10$ in. and produces hard solid prototypes that may be used in many applications.

The FDM technology uses a spool of ABS thermoplastic filament with a diameter of 0.070 in. to build a part. The filament is heated to above its melting point and then extruded through a nozzle on a delivery head and onto a build platform. When one layer is finished the platform is lowered and a new layer is formed on top of it.

The overall inaccuracy of the parts being built by RP technology has been one of the major challenges that needs to be overcome. Errors due to shrinkage and warpage dominate the inaccuracy of the part. The thermoplastic ABS material used in FDM machines experiences a volume change when it is heated and then extruded onto a build platform. "Each prototype made is slightly smaller than its designed dimensions" [12]. SCF is one of the variables that can be controlled by the operator to influence the overall accuracy of the part. When prototyping, the shrinkage is compensated by multiplying the dimensions by an SCF through the build software. Through research, the optimal SCF can be found and the prototype can be built within the specified build accuracy of the machine.

To test the build accuracy of a prototyping machine, a user part, developed by Ed Garguilo of DuPont, which was later adopted by NASUG, was used. It is an in-plane benchmark part that covers the extent of the build platform and allows for the measurement of the part's X and Y dimensions. This particular part (Fig. 11.15) was chosen because of: (1) its large amount of independent surfaces, (2) a variety of balanced measurements ranging from 0.125 to 9.5 in., and (3) an almost equal amount of inside and outside measurements. These three characteristics made the part ideal for performing measurements because it reduced biases on certain measurements, making all of the factors in producing the part just as important [8, 9].

To calculate the SCF, 46 measurements were taken in the X and Y planes [13]. These measurements were input into an Excel program and from these points, 27 outside and 21 inside dimensions were calculated. After obtaining this data, statistical equations were used to determine the overall accuracy of the part.

First the dimensional data is converted to the differences relative to the CAD dimensions. The error in each of the dimensions is calculated by

$$\text{Actual error} = a - c \tag{11.2}$$

where a is the measured dimension and c is the CAD dimension.

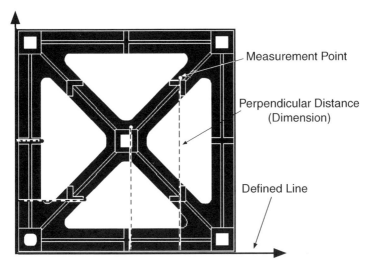

Figure 11.15   NASUG "user part" and the CMM measurement.

Many statistical data analysis computations were derived from the 21 inside dimensions and 27 outside dimensions [13]. The slope of the best-fit regression line of actual error vs. ideal dimension was calculated to get an idea of the accuracy of the part and to determine the new SCF. The slope is determined by

$$\text{Slope} = (42 + 54)*(A + B) - (C - D)*(E - F) \qquad (11.3)$$

where $A = \Sigma(\text{inside ideal dimension})*(\text{error})$
    $B = \Sigma(\text{outside ideal dimension})*(\text{error})$
    $C = \Sigma(\text{outside ideal dimension})$
    $D = \Sigma(\text{inside ideal dimension})$
    $E = \Sigma(\text{error outside dimensions})$
    $F = \Sigma(\text{error inside dimensions})$

The constant is determined by

$$\text{Constant} = (42 + 54)*[\Sigma(G^2) + \Sigma(H^2)] - [\Sigma(H) - \Sigma(K)]*[\Sigma(M) - \Sigma(K)]$$

$$(11.4)$$

where $G = $ inside measured dimension
    $H = $ outside measured dimension
    $K = $ inside ideal dimension
    $M = $ outside ideal dimension

Next a *y*-intercept value must be determined to complete the equation for the least-square regression line for which the slope was found above. This equation is used to find the expected error in each of the dimensions, and, when that value is compared to the actual error, we can determine the residual error in the part. The equation for the intercept is given by

$$\text{Intercept} = [\Sigma(A)^2 + \Sigma(C)^2]*[\Sigma(E) - \Sigma(G)]$$
$$- [\Sigma(I) + \Sigma(L)]*[\Sigma(C) - \Sigma(A)] \qquad (11.5)$$

where I = (outside ideal dimension)*(error)
L = (inside ideal dimension)*(error)
C = outside ideal dimension
A = inside ideal dimension
E = outside error
G = inside error

Now with the slope and intercept defined, the predicted value for the error, given an ideal dimension, can be found by the equation

$$y \text{ (predicted)} = mx + c \qquad (11.6)$$

where m = slope
x = ideal dimension
c = intercept

This process of finding the slope, constant, and intercept is performed for both the *x* and *y* dimensions and the two results are averaged at the end.

The predicted value of error is used to determine the residual square, which squares the difference between the measured dimensions and the predicted value:

$$\text{Residual square} = (\text{difference} - \text{predicted})^2 \qquad (11.7)$$

With the result from Equation 11.7, the residual error in the part can be calculated. The residual error defines the overall difference of the measured values from the predicted error values [10]. The residual error is defined as

$$\text{Residual error} = \sqrt{\frac{(M + N + O + P)}{121}} \qquad (11.8)$$

where M = residual square (inside Y dimensions)
N = residual square (outside Y dimensions)
O = residual square (inside X dimensions)
P = residual square (outside X dimensions)

Finally the SCF can be derived. The SCF is an indication of the numerical percentage of how small or large the user part is compared to the actual CAD dimensions. The SCF equation incorporates the initial shrinkage compensation factor, which was used in the beginning of the build process. The equation to determine the SCF is

$$\text{SCF}(X) = 1 - [1 - (\text{slope})]*[1 + \text{SCF}(X) \text{ OLD}] \qquad (11.9)$$

$$\text{SCF}(Y) = 1 - [1 - (\text{slope})]*[1 + \text{SCF}(Y) \text{ OLD}] \qquad (11.10)$$

11.2.4 Results and Data Analysis

After several trials of making the user part, FDM-1.007* exhibits the lowest mean, rms, and residual error indicating that so far, an SCF of 1.007 gives the lowest error and the most accuracy for the user part.

A scatter graph of the difference between the measured and desired dimension (error) plotted against the desired (CAD) dimension was used in the calculation of the SCF and the total shrinkage (TS) for the user part. The total shrinkage is the slope of a "least-squares" regression line of the error vs. ideal (CAD) dimension graph. It gives an indication of how far off the SCF that was used to create the part is to the SCF that will produce the most accurate part. A negative TS means the part was made too small and a positive TS means the part was made too big. The lowest TS value is desired for the most accurate part. FDM-1.008-P2† had the lowest total shrinkage value, but it was not the most accurate part as indicated by further statistical calculations performed on it. As Figure 11.16 summarizes, the greatest total shrinkage of −0.2966 and 0.2744% was achieved from parts FDM-1.005 and FDM-1.010-II, which indicates that the optimum SCF is not close to those SCF. Part FDM-1.007 showed the second least shrinkage of 0.0453%, but further calculations proved this was the most accurate part.

The SCF is related to total shrinkage because the value for total shrinkage is subtracted from the initial SCF, to get a new, more accurate SCF. As Figure 11.17 illustrates, calculations performed on the measurements revealed that a new SCF that should improve the accuracy for a part made on an FDM machine lies between 0.65 and 0.75%. Part FDM-1.007 was the only part that could be made in this range, and it was this part that turned out to be the most accurate.

The mean error is calculated to get an indication of the error in the part and to determine whether the dimensions were within the machine-specified accuracy [10]. The graph for mean error shows a similar trend to the total shrinkage graph and thus gives similar information. A positive mean error

---

*FDM refers to the build method, 1.007 is the SCF, and I (if used) refers to the first measurement for the part.

†This is the second part made with an SCF of 1.008, due to questionable results given by the first part.

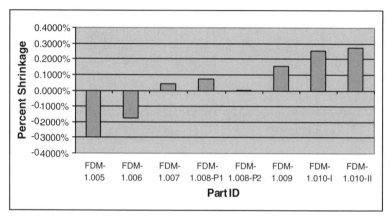

Figure 11.16 Total shrinkage.

indicates that most of the measured dimensions were greater than the CAD dimensions, and a negative mean error indicates that most of the measured dimensions were smaller than the CAD dimensions. The FDM parts with shrinkage factors set from 1.005 to 1.010 had mean errors ranging from −0.0119 to 0.0110 in., giving an even spread about zero. As Figure 11.18 shows, FDM-1.007, FDM-1.008-P1, and FDM-1.008-P2 produced mean errors, which satisfied the machine-specified accuracy of +0.005 in. However, the mean error can be influenced by negative and positive error values and needs to be used in conjunction with other statistical calculations to find the best part.

The standard deviation was also calculated to get an indication of the spread of errors around the mean and really understand the meaning for the

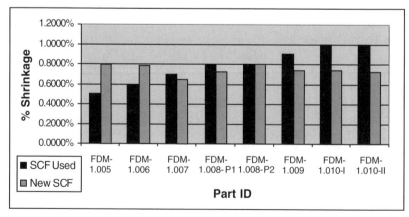

Figure 11.17 Used and new shrinkage compensation factors.

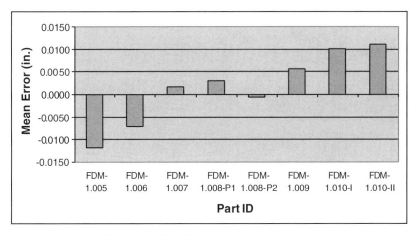

Figure 11.18   Summary of mean error.

value of mean error [10]. A low mean error and a low standard deviation will provide a good indication of the accuracy of a user part. The standard deviation data varied from 0.0096 in. to 0.0246 in. with the full results represented in Figure 11.19. Parts FDM-1.007, FDM-1.008-P1, and FDM-1.008-P2 all had mean errors very close to zero. However, when these values were matched with the standard deviation values, FDM-1.007 had the lowest spread of errors around the mean, and this is another reason for the high accuracy of this part.

Calculating residual error for the user part allows the measurement of the errors in the part, which are not caused by shrinkage. These are random errors in the machine, material, and measurement that are much harder to measure and control. Overall, the standard deviation and residual error (Fig. 11.20)

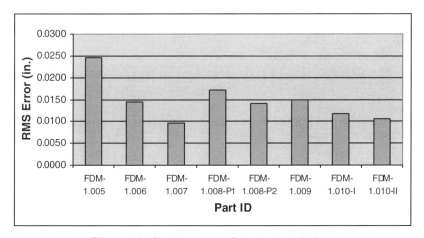

Figure 11.19   Summary of standard deviation.

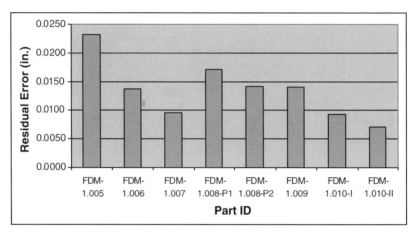

Figure 11.20 Summary of residual error.

for each user part had similar values, indicating that the error in the part due to shrinkage is very small, and it is the random errors that contribute greatly to the inaccuracy of the part. The lowest residual error of 0.0096 in. for FDM-1.007 was the same as its standard deviation, indicating the error in the part due to shrinkage was extremely small and all the errors were random errors that are harder to control.

The most important of all the data calculated is the rms error (Fig. 11.21) as it gives a better indication of the size of the error in each part. It is a better approximation than the mean error, which can be influenced by positive and negative error values [10]. The lowest rms error would indicate the most accurate part (see Table 11.4). It is FDM-1.007, which gives the lowest rms

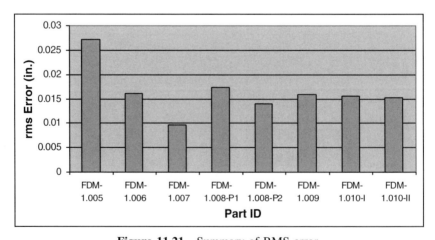

Figure 11.21 Summary of RMS error.

TABLE 11.4 Summary of FDM-1650 Results

Part ID	Slope (%)	SCF Used (%)	New SCF (%)	Mean Error (in.)	SD (in.)	Residual Error (in.)	rms Error (in.)
FDM-1.005	−0.2966	0.5000	0.7981	−0.0119	0.0246	0.0232	0.0272
FDM-1.006	−0.1752	0.6000	0.7800	−0.0071	0.0145	0.0137	0.0160
FDM-1.007	0.0453	0.7000	0.6500	0.0017	0.0096	0.0096	0.0097
FDM-1.008-P1	0.0739	0.8000	0.7253	0.0030	0.0172	0.0172	0.0174
FDM-1.008-P2	0.0006	0.8000	0.7993	−0.0006	0.0142	0.0142	0.0141
FDM-1.009	0.1551	0.9000	0.7442	0.0057	0.0150	0.0141	0.0159
FDM-1.010-I	0.2543	1.0000	0.7400	0.0102	0.0118	0.0093	0.0156
FDM-1.010-II	0.2744	1.0000	0.7200	0.0110	0.0107	0.0071	0.0153

error value indicating its greatest accuracy and verifying the previous results, which indicated that it was the most accurate part. Part FDM-1.008-P2 had lower total shrinkage and mean error values than FDM-1.007, but its standard deviation, root mean square error, and residual error values verified its inaccuracy.

11.2.5 Conclusions and Recommendations

The objective for this research project was to find the optimal shrinkage compensation factor for the FDM-1650 prototyping machine and produce the most accurate part. From the data that was analyzed, the best SCF for the FDM-1650 was SCF = 1.007 or 0.7%.

There are a few recommendations for future accuracy research on the FDM-1650 machine. First, varying temperature and build speed during the build process could affect the accuracy of the parts. These parameters might help minimize the amount of warpage that the part undergoes due to the amount of time it is exposed to heat. Also, building different percent sizes of the part can be used as a test to compare the effects of shrinkage on a small and a large part. Another recommendation is for more accurate SCF inputs, which contain 4 to 5 decimal places. This would allow for accuracy research of parts with SCFs between 1.0065 and 1.0075, the region where the results point to as having the best SCF. The machine should also allow for inputs in both the X and Y direction instead of the one shrinkage factor that can presently be input into the build software. The addition of these inputs and added accuracy in the decimal place should produce a more accurate part that is even closer to its desired dimensions. Finally, the support material that is currently in use with this machine should be changed because it is difficult to remove. A water-soluble support material would make it easier to remove and clean the part, saving valuable time.

Acknowledgment This work was funded by an NSF grant to Loyola Marymount University for its Research Experience for Undergraduates program. The authors also appreciate the support provided by Northrop Grumman and 3D Systems.

11.3 EFFECTS OF CRYOGENIC PROCESSING ON RP MATERIALS

11.3.1 Abstract

Little research has been done on the postprocessing (aging) of rapid prototyped polymers at temperatures below 123 K ($-238°F$). Test specimens of RP thermosetting resin (DSM-Somos 8110) were fabricated and cryogenically aged from 10 to 25 hr. The tensile strength and impact toughness were measured. This work will study the effect of cryogenic aging on yield strength of Somos 8110. This section will also discuss our interpretation of the data based on fractography.

11.3.2 Introduction

As industry continues to increase its use of RP materials, the necessity for enhancing the properties of RP materials is becoming more important. Past studies have concluded that by reducing the temperature below 123 K (−238°F), cryogenic processing increased the strength of many materials, particularly metals [14]. But while the strength of the materials increased, the cryogenic treatments also reduced ductility.

Though relatively unstudied, the results of cryogenic tempering on certain polymers will prove highly valuable to industry. This study hopes to increase the usage of RP materials through cryogenic processing by optimizing the combination of strength and ductility for the stereolithography photopolymer epoxy resin, DSM Somos 8110.

The cryogenic test procedure began with the construction of the tensile and impact test samples using the RP materials. To increase the mechanical properties, these samples were then subjected to 88 K (−300°F) using several different holding and ramping times. Next, the samples were examined using tensile and impact tests. Finally, the data obtained from the tests were analyzed. Based on metals, it was expected that the tensile strength of the RP materials would increase, and the toughness would decrease after cryogenic processing. However, we hoped there would be a combination of cryogenic aging treatments that would improve both the strength and toughness.

The purpose of this study is to discuss the cryogenic method for postprocessing rapid prototyped DSM-Somos 8110 epoxy resin. In addition, we will examine the effect of cryogenic aging on tensile strength and impact energy.

11.3.3 Background

Cryogenics Cryogenics is the science and art of producing a low-temperature environment. It started in 1877 when two scientists, Cailletet in Paris and Pictet in Geneva, developed a procedure to liquefy oxygen in the laboratory [14]. Now, liquid nitrogen is one of the most common cooling media. Since the normal boiling points of nitrogen and other permanent gases such as helium, oxygen, and argon is about 84.8 K (≈−307°F), the cryogenic temperature is generally considered to be 84.8 K or below [15].

Cryogenic processing is an important field in industry today. For example, an increase of 195 to 817% in wear resistance has been reported for standard steel that was cryogenically treated [16]. The cryogenic process consists of three stages that are time and temperature dependent to the aging temperature. This process starts with the gradual ramping down of temperature for a predetermined time range to a specific point. The temperature is then held for a period of time. Finally, a gradual process of temperature ramp-up is executed. As a result of cryogenic processing, some unknown molecular changes occur, modifying the structure of the material [16].

Conventionally, heat treating is used to strengthen steel and other alloys. Upon quenching steel below its eutectoid decomposition temperature, some retained austenite remains. However, cryogenic processing transforms virtually all of the retained austenite to martensite, along with a precipitation of fine carbide particles. These effects have resulted in increased wear resistance [16].

Polymers and Stresses   Polymers can be classified according to their chain structure and chain arrangement. Chain structure of polymers is further characterized by the chemical composition of monomers and their relative arrangement within a chain. The common chain arrangements are linear chains, branched chains, and cross-linked networks (e.g., epoxy resins).

The general characterization of polymers is very difficult due to the complicated structures, the great variety of chemical compositions, and the transfer of molecular groups [16]. However, the fracture strength of polymers can be explained from a microstructural point of view. These factors can include molecular stress concentrations and residual thermal stresses [17].

The twisting and kinking of polymer chains contributes to local stress concentrations due to the uneven stress distribution between the molecules. When the entangled polymer chains are under stress, the chain segments in the load direction provide most of the load-carrying capacity. When polymers are cooled too quickly during cryogenic processing, thermal stresses are built up within the polymer, and these act to reduce the tensile strength.

11.3.4   Experimental Procedure

Equipment and Process   The following major equipment was used in this research.

1. Northrop Grumman SLA-250 RP machine
2. Cryogenic treatment equipment
3. Instron tensile testing machine
4. Izod impact tester
5. Scanning electron microscope (SEM)

The experimental setup if shown in Figure 11.22.

The following outlines the steps used in the experimental process that was used to design, fabricate, test, and analyze the samples.

1. Design dog bone and v-notch samples
2. RP machine builds parts from design

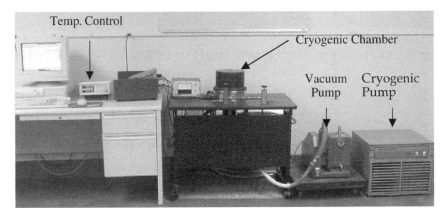

Figure 11.22   Experimental setup for cryogenic processing.

3. Expose parts to cryogenic treatment with liquid He
4. Tensile and impact testing
5. SEM fractography

Designing and Prototyping Samples   Drawings of the dog bone and v-notch-shaped samples were created using AutoCAD (Figs. 11.23*a* and 11.23*b*), and these were saved as separate .DWG files. These files were then converted into STL format for use with stereolithography software. SLA 250 (Northrop Grumman) RP machines were then used to rapid prototype the parts.

Cryogenic Treatment   The following process was used to cryogenically treat each sample.

1. All samples except the baseline went through cryogenic treatment before testing. The samples were prepared at Northrop Grumman and cryogenically aged at Loyola Marymount University.
2. The cryogenic process is characterized by three parameters: ramp-down time from room temperature to 88 K ($-300°F$), hold time at 88 K, and ramp-up time from 88 K to room temperature.
3. Preliminary experiments were performed on cryogenic treatment. The ramp-down time of 14 hr was used. Three holding times of 10, 15, and 20 hr, and ramp-up times of 14 hr were used.
4. Samples are labeled as follows: XX-XX-XX. The numbers represent the ramp-down time, holding time, and ramp-up time, respectively, in hours (e.g., 14-10-14 means that the samples were ramped-down in 14 hr, cryogenically aged 10 hr, and then ramped-up in 14 hr.)

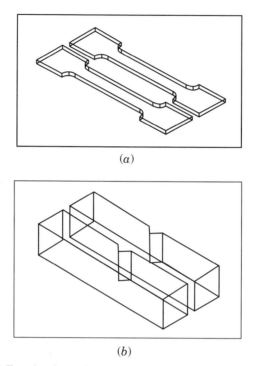

Figure 11.23   (*a*) Two dog-bone-shaped samples and (*b*) two v-notch-shaped samples.

Tensile Testing   The yield and ultimate strengths of the samples were measured using the Instron Universal Testing Instrument 4500. The cross-head speed of the test machine was 0.0212 mm/sec. The interface with the machine was performed using the front panel and a software program running on a desktop computer. The test specimens were designed in the shape of a dog bone, similar to ASTM D638-97. The end pads of the specimen were tightened and aligned between the two grips of the frame unit. A computerized load cell located inside the frame unit measured the force applied to the gauge section of the sample. The software calculated the tensile stress in the gauge section. Strain was measured with an extensometer, and the stress vs. strain curves were plotted as the specimen was continuously loaded. The yield stress was measured at 0.2% offset strain, and the ultimate strength was measured at the maximum stress the material could withstand. Four to six specimens were tested for each cryogenic aging treatment (0 to 20 hr). The tensile strength data were statistically analyzed to determine the effect of aging treatment on the yield and ultimate strengths.

Izod Impact Testing   Izod impact testing is performed to determine the toughness of a material. The sample is made with a centered v-shaped notch. During the impact testing, the sample is subjected to a quick and intense blow by a hammer pendulum. The impact test evaluates the material's resistance to crack propagation. The impact energy absorbed by the sample during failure is determined by calculating the difference in potential energy of the hammer.

SEM Fractography   The fracture surfaces of the failed tensile samples of Somos 8110 resin were examined under a scanning electron microscope. SEM fractography was used to interpret the results of the tensile strength data for both the untreated (baseline) and cryogenically treated samples [18]. The SEM used an acceleration voltage of 1.2 kV. The fracture surfaces were observed sequentially at magnifications of 15×, 40×, and 150×.

The stretching of the polymer chains and fracture initiation were correlated with the yield strength data. Here critical-size cracks were initiated from defects and propagated when the yield strength of the polymer was exceeded. The propagation of the crack front was correlated with the ultimate strength. Then the crack propagation patterns could be observed on the fracture surface of the failed specimens.

11.3.5   Results and Discussion

Tensile and Impact Testing   The results of the yield strength and ultimate strength vs. cryogenic treatment time (0 to 20 hr) are shown in Figures 11.24*a* and 11.24*b*. In both cases, the strengths appear to be affected by aging time. To verify these results, the data were statistically analyzed using multiple *t* testing, which compared the means of two treatments at a time at a 0.05 level of significance [19]. A one-way analysis of variance (ANOVA) could not be used because the variances of the treatments were unequal. For the yield strength data (Fig. 11.24*a*), it was determined with 95% confidence that the means of the treatments were statistically insignificant. Due to the large variances at 15 and 20 hr, we can say that there was no significant increase in the yield strength with aging time.

For the ultimate strength (Fig. 11.24*b*) vs. aging time, the data were analyzed in the same way. Again, one-way ANOVA could be not be used. Multiple *t* testing on the treatment means showed that for 95% confidence two conclusions were reached: (1) the mean strengths at 0, 10, and 20 hr were equivalent, and (2) the mean strength at 15 hr was significantly lower than that at 10 and 20 hours. Hence, aging time produced a drop in the ultimate strength between 10 and 15 hr. Otherwise, there was no significant effect of cryogenic aging on the ultimate stress.

When the impact energy data vs. aging time (Fig. 11.24*c*) were statistically analyzed, the large variance at 10 hr indicated that the treatment means were insignificant with 95% confidence. Hence, there was no effect of cryogenic treatment time on the impact energy.

SEM Fractography   Since all of the tensile specimens failed near the fillet radius, failure analysis was performed on the tested specimens. The fracture surfaces of the tensile specimens were examined under three different treat-

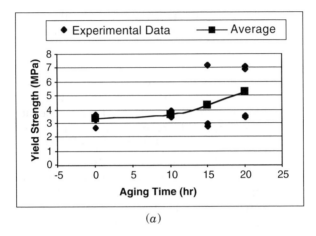

(a)

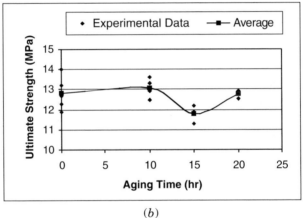

(b)

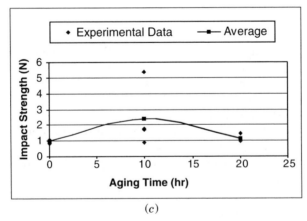

(c)

Figure 11.24 (a) Yield strength vs. cryogenic aging time, (b) ultimate strength vs. cryogenic aging time, (c) impact energy vs. cryogenic aging time.

ment conditions: (1) sample A, treated for 15 hr and having the *highest* yield strength (Fig. 11.24*a*), (2) sample B, treated for 15 hr and having the *lowest* yield strength (Fig. 11.24*a*), and (3) sample C, untreated.

The following findings were made from observing the fracture surfaces of samples A, B, and C in Figures 11.25*a*, 11.25*b*, and 11.25*c*, respectively. In all cases, the fracture originated from surface defects on the fillet radius (between the gauge length and grips) of the tensile specimens. Fracture propagated from left to right, as shown by the "river markings" on the fracture surface [18]. Samples B and C showed similar fracture patterns that consisted of mixed mode I and III fractures [18]. Here, fracture initiated from the stepped fillet radius in Figures 11.25*b*, 11.25*c*, and 11.26*b*, which was formed during stereolithographic processing. Sample A exhibited primarily mode I fracture from the fillet surface, and fracture did not originate from steps on the fillet radius (Figs. 11.25*a* and 11.26*a*).

Samples B and C had low yield strengths and similar fracture patterns. The low strengths were interpreted as being caused by the stepped fillet radius. The steps probably caused a high stress concentration at the fillet surface. In addition, the steps appeared to have surface defects on the fillet radius (Figs. 11.25*b*, 11.25*c*, and 11.26*b*). Sample A had the highest yield strength, and there were no steps or surface defects on the fillet radius where fracture originated. Hence, the stepped fillet surfaces with their associated surface defects probably caused the specimens to fail prematurely at lower strengths. The steps and surface defects in samples B and C also accounted for the higher variance (scatter) in the data.

The differences in the fracture patterns between sample A and samples B and C could have also accounted for their difference in yield strength. In samples B and C, it appeared that the crack front propagated at slower velocities than in sample A. The crack had more time to separate on to different levels and develop mode III shear lips (Figs. 11.25*b* and 11.25*c*). In sample A, there were no steps and surface defects at the fillet radius. Consequently, a critical-size crack had to be nucleated on the fillet surface, and this required extra stress, which caused the strength to be higher. Once a critical-size crack was formed, the extra stress caused the crack to propagate catastrophically at a higher velocity. The higher velocity fracture in sample A is indicative primarily of mode I fracture patterns with little mode III shearing (Fig. 11.25*a*).

11.3.6 Conclusions and Recommendations

Based on the findings from our research, the following conclusions and recommendations can be made:

1. DSL-Somos 8110 test specimens were fabricated by stereolithography using SLA 250 RP equipment. The specimens were laser cured by pulling the specimens parallel to the tensile axis.

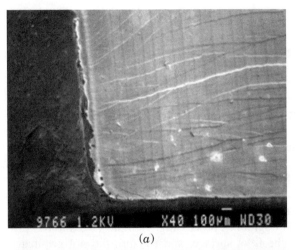

(a)

(b)

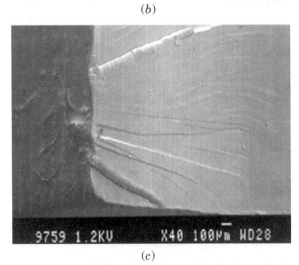

(c)

Figure 11.25 (a) 40× SEM micrograph of sample A in Figure 11.24(a), (b) 40× SEM micrograph of sample B in Figure 11.24(a), and (c) 40× SEM micrograph of sample C (baseline) in Figure 11.24(a).

(a)

(b)

Figure 11.26 (*a*) 15× SEM micrograph of sample A from Figure 11.25(*a*) and (*b*) 15× SEM micrograph of sample C (baseline) from Figure 11.25(*c*).

2. The test specimens were cryogenically treated by a ramp-down cycle from room temperature to 88 K in 14 hr, a hold cycle of 10 to 20 hr, and then a ramp-up cycle from 88 K to room temperature in 14 hr.

3. Due to large data scatter, the yield strength and impact energy were not affected by cryogenic aging treatment. Only the ultimate strength exhibited a significant decrease with aging treatment from 10 to 15 hr.

4. It was expected the wide data scatter was caused by microscopic steps and their associated defects at the fillet radius of the tensile specimens. Samples with steps and defects had lower strengths, and those without

steps and flaws had high strengths. The variability in the surface texture of the fillet radius caused a wide variability in the tensile strength and impact energy.

5. For future work, it is recommended that the SLA specimens be rapidly prototyped normal to the face of the dog-bone specimen. This will hopefully reduce the data scatter, which currently masks any potential effect of cryogenic aging time on the tensile strength and impact energy.

Acknowledgments   The work was funded by an NSF grant under Research Experiences for Undergraduates (REU). The authors also appreciate the support provided by the Northrop Grumman Corporation. We also acknowledge the contributions toward the development of this study by our students, J. C. Jackson and Joseph A. Weinmann.

11.4   NEW TECHNOLOGIES

Many RP research and developments are taking place at universities, industries, and government laboratories around the world. The following technologies are expected to impact the field of RP.

11.4.1   Laser Engineered Net Shaping

Laser engineered net shaping (LENS) is one of the first direct metal rapid prototyping systems (see Fig. 11.27). The parts produced are full-strength metals upon removing them from the machine. Sandia National Laboratories

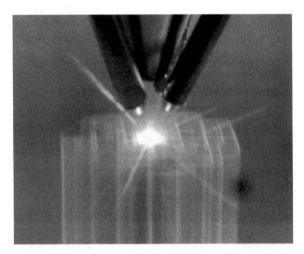

Figure 11.27   Laser engineered net shaping (LENS) rapid prototyping system. (Courtesy of Sandia National Lab.)

developed the LENS process in conjunction with various industry partners on a Cooperative Research and Development Agreement (CRADA) [20].

Currently, the LENS process uses metals such as stainless steel 316 (SS316), tooling steel (H13), and titanium with 6% aluminum and 4% vanadium (Ti-6-4). The LENS system continues to incorporate widely used RP techniques such as the layer-by-layer process. This process takes a CAD model and creates an STL file, which is then downloaded to the machine where it will be produced.

Once the machine is prepared, four individual nozzles direct streams of metal powders at a central point beneath, where a high-powered laser beam heats the metal. The laser and jets remain stationary while the model and its substrates are moved to continually provide new targets on which to deposit metal. An argon environment is created in order to prevent oxidation of the powders during the build process [21].

11.4.2 Solidica

Solidica, Inc., located in Ann Arbor, Michigan, has incorporated traditional CNC milling with ultrasonic welding to produce direct metal fabrication technology. The Formation 2030 uses a layer-by-layer process and ultrasonic welding to consolidate aluminum alloy tape that is 25 mm wide by 0.1 mm thick and CNC milling to shape each layer (see Fig. 11.28). Ultrasonic welding technology uses sound to create microfriction between each layer of metal tape. Under pressure, the metal layers are caused to form metallurgical bonds.

Figure 11.28 Solidica Formation 2030 rapid prototyping system. (Courtesy of Solidica.)

Ultrasonic consolidation produces parts with 98 to 99% with no discernable boundaries between layers.

11.4.3 On Demand Manufacturing

On Demand Manufacturing (ODM), a new subsidiary of the Boeing Company (June 2002), has purchased an initial two Vanguard selective laser sintering systems for use in the rapid fabrication of complex, hard-to-manufacture parts (see Fig. 11.29). In the past, prototype services have made small runs of production or replacement parts using rapid prototyping technologies. But ODM is the first company to eschew prototype manufacturing and instead focus on short-run production parts for industrial applications. ODM's first contract will be to supply Boeing with parts for the air-conditioning ducts of military aircraft.

11.4.4 Direct Material Deposition: OPTOMEC

Optomec is an Albuquerque-based supplier of laser-directed material deposition systems that markets LENS technology. Optomec delivers unique additive manufacturing solutions for high-performance applications in defense, aerospace, electronics, medical, and industrial markets. Their commercial LENS 850 systems enable the cost-effective development, production, and service of a wide range of end products (see Fig. 11.30). The LENS 850

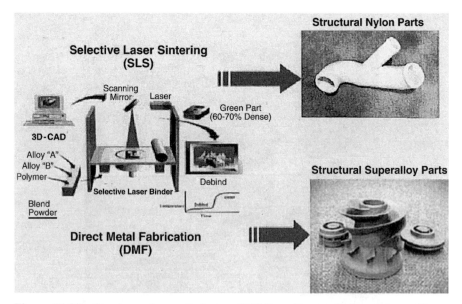

Figure 11.29 On–demand manufacturing (ODM) rapid prototyping systems. (Courtesy of Optomec.)

Figure 11.30 Optomec 850 laser engineered net shaping (LENS) rapid prototyping system. (Courtesy of Optomec.)

system with its 7.5 cubic feet build envelope is designed to handle most components. In addition, dual power feeders with gradient control and an upgraded 1000W Nd:YAG laser are standard with this machine. This comes with all of the other standard benefits of a LENS system.

Optomec's LENS technology provides real art-to-part capability for rapid manufacture of fully dense metal parts. Optomec's proprietary software takes STL file and converts it to motion control file ready for the LENS system. The fabrication process starts with high power Nd:YAG laser focusing onto metal substrate to create a molten puddle on the substrate surface. The metal powder is then injected into the molten puddle on the substrate to increase the metal volume. The deposition head is scanned relative to the substrate to write lines of the metal with a finite width and thickness, forming layers to create the final near net shape part (see Figure 11.31). The key features of the LENS process are the small heat affected zone (HAZ) and the rapid cooling of the weld pool. The small HAZ minimizes the impact to the substrate. The rapid cooling creates very fine microstructural feature, delivering high tensile strength and high ductility for deposited metal. The primary application for LENS technology include repair and overhaul, rapid prototyping, low-volume manufacturing and product development for aerospace, defense, and medical markets.

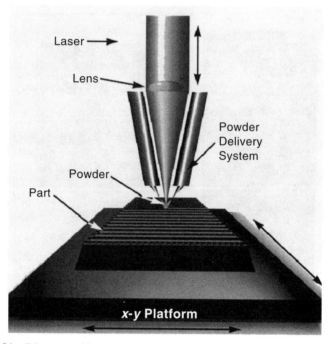

Figure 11.31   Diagram of laser engineered net shaping (LENS). (Courtesy of Optomec.)

11.5   SUMMARY

This chapter presents examples of research and development efforts of the author and his students at LMU by way of illustration of the type of research that is both useful to industry and academia. However, there are many excellent research and development projects being performed in both academia and industry, and it is left entirely to the instructor to pick and choose the most appropriate and timely examples available to illustrate relevant concepts at the time the course is presented.

The first examples deals with the improvement of FDM process using design of experiments technique. A two-level, four-factor, full factorial design was used for the project. The project detals the design matrix, preparation of samples, testing, and analysis of samples. The project describes the physical interpretation of the data also.

The second example investigates the present in-plane accuracy of a particular FDM machine using the benchmark "user part" and shows the effect of optimal shrinkage compensation factors (SCF) on the accuracy of the prototyped parts. The benchmark parts were built on the FDM-1650 prototyping machine and a total of 46 measurements were taken in the X and Y planes using a Brown & Sharpe coordinate measuring machine (CMM). The data was then analyzed for accuracy using standard formulas and statistics, such as mean error, standard deviation, residual error, rms error, and so forth. The optimal SCF for the FDM-1650 machine was found to be 1.007, or 0.7%.

The third project investigates the effects of cryogenic processing on the tensile and ultimate strengths of prototyped samples. Test specimens of RP thermosetting resin (DSM Somos-8110) were fabricated and cryogenically aged from 10 to 25 hr. The tensile strengths and impact roughness were measured. The research discusses the interpretation of the data based on fractography.

The last section of the chapter briefly describes some new technologies such as laser engineered net shaping (LENS), Solidica, On Demand Manufacturing (ODM), and direct material deposition (DMD) by Optomec. These technologies are the results of many RP research and developments that are taking place in industry and academia and in some government laboratories.

PROBLEMS

11.1. Consider the two-level, eight-trial matrix for the 2^3 classical (Western) and the L_8 Taguchi experimental design, as shown below. Assume that $+1$ (high, classical) $=$ 2 (high, Taguchi) and -1 (low, classical) $=1$ (low, Taguchi).

a. Are the two experimental designs equivalent? Please explain.

b. Prove that the classical 2^3 design is orthogonal.

Classical 2^3 Design							Taguchi L_8 Design						
A	B	C	AB	AC	BC	ABC	1	2	3	4	5	6	7
-1	-1	-1	$+1$	$+1$	$+1$	-1	1	1	1	1	1	1	1
$+1$	-1	-1	-1	-1	$+1$	$+1$	1	1	1	2	2	2	2
-1	$+1$	-1	-1	$+1$	-1	$+1$	1	2	2	1	1	2	2
$+1$	$+1$	-1	$+1$	-1	-1	-1	1	2	2	2	2	1	1
-1	-1	$+1$	$+1$	-1	-1	$+1$	2	1	2	1	2	1	2
$+1$	-1	$+1$	-1	$+1$	-1	-1	2	1	2	2	1	2	1
-1	$+1$	$+1$	-1	-1	$+1$	-1	2	2	1	1	2	2	1
$+1$	$+1$	$+1$	$+1$	$+1$	$+1$	$+1$	2	2	1	2	1	1	2

11.2. An injection-molding process produces thick plastic panels. The problem is the panels have a flatness variation that depends on the processing parameters. The degree of flatness was chosen to be the response function. The goal of the experiment was to minimize the flatness. The design team concluded that four factors affected the flatness: melt temperature, mold temperature, cure time, and injection speed. The factors and their levels are shown below.

FACTORS VS. LEVELS

Factor	Low (1) Level	High (2) Level
A. Melt temperature	500 K	550 K
B. Mold temperature	80 K	140 K
C. Cure time	150 sec	200 sec
D. Injection speed	1.00 mm/sec	2.25 mm/sec

An eight-trial, two-level, four-factor fractional-factorial experiment was performed. Using the L_8 experimental design matrix below, the flatness data was measured randomly by the Y_i values for all experimental trials.

Random	Trial	A	B	C	AB	AC	BC	D	Y_i
2	1	1	1	1	2	2	2	1	54
5	2	1	1	2	2	1	1	2	55
1	3	1	2	1	1	2	1	2	46
4	4	1	2	2	1	1	2	1	50
7	5	2	1	1	1	2	2	2	46
3	6	2	1	2	1	1	1	1	45
6	7	2	2	1	2	2	1	1	30
8	8	2	2	2	2	1	2	2	24

a. Construct a response table and determine the effects of the factors and interactions AB, AC, and BC.

b. Plot the effects on a normal probability graph, and state your conclusions.

c. Graph the response Y_{min} vs. the effects.

d. Prepare a separate graph on the effect of the AB interaction on the response.

e. Write the response equation. How should it be modified?

f. What conclusion can you draw from the effects of the factors and interactions on the response?

REFERENCES

1. Stratasys, http://www.stratasys.com.

2. O. Es-Said, R. Noorani, M. Mendelson, J. Foyos, and R. Marloth. Effect of Layer Orientation on Mechanical Properties of Rapid Prototyped Samples, *Materials and Manufacturing Processes,* 15(1), 107–122, 2000.

3. M. Montero, S. Roundy, D. Odell, S-H. Ahn, and P. Wright. Material Characterization of Fused Deposition Modeling (FDM) ABS by Designed Experiments, Society of Manufacturing Engineers, Annual Conference, Cincinnati, OH, 2001, pp. 1–21.

4. J. Walish, M. Sutaria, M. Dougherty, R. Vaidyanathan, S. Kasichainula, and P. Calvert. Application of Design of Experiments to Extrusion Freeform Fabrication (EFF) of Functional Ceramic Prototypes, *Solid Freeform Fabrication Conference,* Proceedings, Univ. of Texas, Austin, 1999, pp. 103–110.

5. D. C. Montgomery. *Design and Analysis of Experiments,* 5th ed., Wiley, New York, 2001.

6. R. Noorani, P. Gerencher, M. Mendelson, O. S. Es-Said, and S. Dorman. Impact of Rapid Prototyping on Product Development, Proceedings of the International Conference on Manufacturing Automation, Hong Kong, April 28–30, 1997.

7. E. P. Gargiulo, (E.I. DuPOnt de Nemours and Co., Wilmington, DE) and D. A. Belfiore (Pennsylvania State University, State College, PA), *Stereolithography Process Accuracy User Experience,* 1995.

8. R. Noorani, Q. Dao, J. Frimodig, H. Le, X. Li, and S. Putnam. Calculation of Shrinkage Compensation Factors for Rapid Prototyping. *Computer Applications in Engineering Education,* **7**(3), 1999.

9. E. P. Gargiulo, and D. A. Belfiore. "Photopolymer Solid Imaging Process Accuracy," *Intelligent Design and Manufacturing for Prototyping,* ASME PED, **50,** 1991.

10. P. Jacobs. *Stereolithography and other RP&M Technologies,* Society of Manufacturing Engineers, Dearborn, Michigan, 1996.

11. Stratasys, Inc., *FDM System Documentation,* E. P. Gargiulo, (Dade International Inc.) and S. S. Jayanthi, (Dupont Company), *Current State of Accuracy in Stereolithography,* 1996.

12. Stratasys, *FDM System Documentation,* F1650-5, F1650-6, F1650-11, G-8, G-9, and Q79, 1996.

13. D. Freedman, R. Pisani, and R. Purves. *Statistics,* 2nd ed., W. W. Norton, New York, 1991.

14. R. G. Scurlock. *History and Origins of Cryogenics,* Institute of Cryogenics, University of Southampton, UK, Clarendon Press, Oxford, 1992.

15. V. Chhabra. Ingenious Cryogenics: Gainsville Inventor Frank Masyada Brings the Future to Our Fingertips, *Business,* 1994.

16. G. Hartwig. *Polymer Properties at Room and Cryogenic Temperatures,* Plenum Press, New York and London, 1995.

17. D. R. Askeland. *The Science and Engineering of Materials,* 3rd ed., PWS Publishing, Boston, 1995.

18. D. Hull. *Fractography: Observing, Measuring and Interpreting Fracture Surface Topography,* Cambridge University Press, Cambridge, UK, 1999.

19. D. C. Montgomery, G. C. Runger, and N. R. Nubele. *Engineering Statistics,* 2nd ed., Wiley, New York, 2000.

20. Sandia Labs, *Intense Hopes: Ten Companies Team to Commercialize Sandia's Powder-to-Parts Net Shaping Technology,* 1948, *http://www.sandia.gov/LabNews/LN12-05-97/lens_story.html.*

21. K. Cooper. *Rapid Prototyping Technology: Selection and Application,* Marcel Dekker, New York, 2001.

APPENDIX A

RP RESOURCES

In doing research for this book, many resources were found and used. Following are a select group of resources that provides the most complete and up-to-date research and development information on rapid prototyping. It should be noted that resources relating to each specific chapter were provided in the reference section at the end of each chapter.

1. *Wohler's Associates Inc.* (*www.wohlersassociates.com*)   In addition to the *Wohler's Report,* which provides information of annual worldwide progress in the rapid prototyping and tooling industries, much information can be found by going online to the website shown above. The site includes additional information on the subjects in the written report, links to hundreds of articles, technical papers, reports, and other documents related to RP, RT, 3D tooling, and reverse engineering. The website also provides links to RP system manufacturers, CAD vendors, service providers, universities, and other organizations affiliated with RP development.

2. *Clemson University* (*http://www.vr.clemson.edu/rp/rp_lierature.htm*) The Clemson University website is an extremely useful tool for learning more about RP-related topics such as specifications, academia, literature, RP systems, services, software, rapid tooling, reverse engineering, and other RP-related topics. Through the website, users can visit many of the research centers and programs available at Clemson. Among these research centers is the Laboratory to Advance Industrial Prototyping (LAIP). Falling within the Center for Advanced Manufacturing in the College of Engineering and Science, the laboratory supports the university's mission in research and public service by providing technology transfer to public and private sectors.

3. *University of Utah* (*http://www.cc.utah.edu/~asn8200/rapid.html*) The Rapid Prototyping Home Page at the University of Utah is one of the most extensive and informative sites available. The website provides links to many commercial RP systems, concept modelers, resellers of concept modelers and RP systems, consultants, and literature and conferences to name a few. In addition, there is an extensive list of commercial service providers and a list of the foremost research going on in academia. The website is truly centered around letting the user access as much information as possible in almost all RP-related fields.

4. *Internet Mail List*   A free rapid prototyping mail list is available to anyone with an e-mail address. In order to subscribe, an e-mail message has to be sent to *majordomo@rapid.lpt.fi* in which the words *subscribe rp-ml* must be entered in the body of the message. Once you are a subscriber, you can send in questions and comments that will be received by more that 1700 companies worldwide.

5. *Global Alliance of Rapid Prototyping Associations* (*GARPA*)   This association was formed to encourage members to exchange information about RP-related issues across international borders. Members of GARPA include associations and organizations in Australia, Canada, China, Germany, Hong Kong, Italy, the United Kingdom, and the United States to name a few. Information on how to become a member of these groups or associations is found at *www.garpa.org*.

6. *RPA/SME*   The Rapid Prototyping Association of the Society of Manufacturing Engineers (RPA/SME), is the world's largest individual member association on RP. This association focuses on the technologies and processes that help develop, test, revise, and manufacture new products in order to bring them into the market faster and cheaper. For more information on RPA/SME visit the website *www.sme.org/rpa*.

7. Worldwide Guide to Rapid Prototyping, Castle Island (*home.att.net/~castleisland*).

APPENDIX B*

WORLDWIDE RP
SYSTEM MANUFACTURERS

*Adapted from company websites and Wohlers Reports 2002, 2004.

Manufacturer	Location	Contacts	Website
		United States	
3D Systems, Inc.	26081 Avenue Hall Valencia, CA. 91355	888-337-9786 (toll free in the U.S.) 661-295-5600 Fax 661-294-8406	*www.3dsystems.com*
Cubic Technologies, Inc.	1000 E. Dominguez Street Carson, CA. 90746-3608	310-965-0006 Fax 310-965-0141 *meygin@cubictechnologies*	*www.cubictechnologies.com*
Optomec, Inc.	3911 Singer Blvd., NE Albuquerque, NM 87109	505-761-6638 Fax 505-761-6638 *info@optomec.com*	*www.optomec.com*
ProMetal: A Division of Extrude Hone	Am Mittleren Moos, 41 D-86167 Augsburg Germany	+49 (0)821 7476 100 Fax +49 (0)821 7476 111 *info@prometal-rct.de*	*www.prometal-rt.com*
Sanders Design International, Inc.	P.O. Box 550 Wilton, NH 03086	603-654-6100	*www.sandersdesign.com*
Schroff Development Corp.	P.O. Box 1334 Mission, KS 66222	913-262-2664 Fax 913-722-4936 *info@schroff.com*	*www.schroff.com*
Solidica	3941 Research Park Drive Suite C Ann Arbor, MI 48108	734-222-4680 Fax 734-222-4681 *info@solidica.com*	*www.solidica.com*
Solidscape, Inc.	316 Daniel Webster Highway Merrimack, NH 03054-4115	603-429-9700 Fax 603-424-1850 *precision@solid-scape.com*	*www.solid-scape.com*
Stratasys Inc.	14950 Martin Drive Eden Prairie, MN 55344-2020	800-937-3010 (toll free in the U.S.) 952-937-3000 Fax 952-937-0070 *info@stratasys.com*	*www.stratasys.com*
Z Corporation	32 Second Avenue Burlington, MA. 01803	781-852-5005 Fax 781-852-5100 *sales@zcorp.com*	*www.zcorp.com*

Manufacturer	Location	Contacts	Website
		Israel	
Objet Geometries Ltd.	Kiryat Weizmann Science Park P.O. Box 2496 Rehovot, 76124	+972-8-931-4314 Fax +972-8-931-4315 info@2objet.com	www.objet.co.il
Solidimension	Shraga Katz Bldg. Be'erot Itzhak 60905	972-3-9033-190 Fax 972-3-9037-304	www.solidimension.com
		Europe	
Envision Technologies GmbH	Blbestraβe 51 D-45768 Marl Germany	49-(0)2043-98750 Fax 49-(0)2043-987599 hendrik.john@envisiontec.de	www.envisiontec.de
EOS GmbH Electro Optical Systems	Pasinger Strasse 2 D-82152 Planegg/Munich Germany	49-89-856-36-0 Fax 49-89-856-36-285 info@eos-gmbh.de	www.eos-gmbh.de
F&S GmbH	Technologie-Park 12 D-33100 Paderborn Germany	49-5251-6-22-55 Fax 49-5251-6-30-62 info@fockeleundschwarze.de	www.fockeleundschwarze.de
Charlyrobot	B.P. 22 F74350 Cernex France	33-4-50-32-80-00 Fax 33-4-50-44-00-41 info@charlyrobot.com	www.charlyrobot.com
Autostrade Co., Ltd.	31-54 Ueno-machi Oita-City Oita 087-0832	81-97-543-1491 Fax 81-97-545-3910 rps@autostrade.co.jp	www.autostrade.co.jp
CMET Inc.	Sumitomo Fudosan Shin-Yokohama Bldg. 2-5-5, Shin-Yokohama Kouhoku-ku, Yokohama, Kanagawa 222-0033 Japan	+81-45-478-5561 Fax +81-45-478-5569 sales@cmet.co.jp	www.cmet.co.jp

Denken Engineering Co., Ltd	2-1-40, Sekiden-machi Oita City, Oita Pref., 870 Japan	81-97-583-5535 Fax 81-97-583-5580 hayashi@denken-eng.co.jp	www.oitaweb.ne.jp/dksystem
Sony/D-MEC Ltd.	JSR Building 2-11-24, Tsukiji Chuo-ku, Tokyo 104-0045	81-3-5565-6661 Fax 81-3-5565-6641 tokyo@d-mec.co.jp	www.d-mec.co.jp
KIRA Corporation	Tomiyoshi-shinden Kira-cho, Hazu-gun	0563-32-1161 Fax 0563-32-3241	www.kiracorp.co.jp
KIRA American Corporation	4133 Courtney Street Unit 3 Franksville, WI 53126	414-835-9272 Fax 414-835-9273	
Toyoda Machine Works., Ltd.	Post Code 448-8652 1, Asahimachi 1-chrome	81-566-25-5171 Fax 81-566-25-5470	www.toyoda-kouki.co.jp
Unirapid Inc.	1-5-2-703 Satsukidaira Misato-shi Saitama-ken 341-0021	81-48-950-1460 Fax 81-48-950-1461	www.webs.to/uri

| Beijing Yinhua Laser Rapid Prototypes Making and Mould Technology Co. Ltd. | Department of Mechanical Engineering Tsinghua University Beijing 100084 | 8610-62783565 Fax 8610-62785718 dmeyyn@mail.tsinhua.edu.cn | |

| Kinergy PTE Ltd. | Block 5002, #03-08, TECHplace II Ang Mo Kio Avenue 5 | (65) 6481 0211 Fax (65) 6482 0179 BuzDev@kinergy.com.sg | www.kinergy.com.sg |

357

APPENDIX C

RAPID TOOLING TECHNOLOGY SUPPLIERS

Manufacturer	Location	Contacts	Website
		United States	
3D Systems, Inc.	26081 Avenue Hall Valencia, CA. 91355	888-337-9786 (toll free in the U.S.) 661-295-5600 Fax 661-294-8406	www.3dsystems.com
APPEX	Stuttgarter-Str. 7 D-80807 Munchen	089-3 58 70-600 Fax 089-3 58 70-798 info@appex.com	
BASTECH INC	849 Scholz Dr Vandalia, OH 45377	937-890-9292 Fax 937-890-9293	bastech@bastech.com
Dynamic Tool Company	1421 Vanderbuilt Dr. El Paso, TX 79935	915-598-2330 Fax 915-298-7150	www.dynamictool.com
Ford Sprayform	Patent and Technology Licensing Office	313-594-1993 Fax 313-323-2647 hfradkin@ford.com	
Optomec, Inc.	3911 Singer Blvd. NE Albuquerque, NM 87109	505-761-6638 Fax 505-761-6638 pferguson@optomec.com	
RSP Tooling	30555 Solon Industrial Pkwy Solon, OH 44139-4329	440-505-6033 Fax 440-248-3882	www.rsptooling.com
Paramount Industries, Inc.	2475 Big Oak Road Langhorne, PA 19047	215-757-9611 Fax 215-757-9784	sales@paramountind.com
ProMetal: The Ex One Company	P.O. Box 1111 Irwin, PA 15642	800-367-1109 (toll free in the U.S.) Fax 724-863-8759 prometal@extrudehone.com	www.prometal-rt.com
RePilForm	1583 Sulphur Spring Road, Suite 104 Baltimore, MD 21227	410-242-5110 Fax 530-325-4768 info@rapid-tech.co.uk	www.info@repliforminc.com

359

Manufacturer	Location	Contacts	Website
		United States	
RPDG (Rapid Product Development Group)	2565 Summit Drive San Deigo, CA 92025	760-703-5770 info@rpdg.com	www.hr@rpdg.com
Solidica	3941 Research Park Drive, Suite C Ann Arbor, MI 48108	734-222-4680 info@solidica.com	www.solidica.com
		Europe	
Arcam Amsterdam	Prins Hendrikkade 600 1011 VX Amsterdam	020-620-48-78 Fax 020-638-55-98 earcam@arcam.nl	www.arcam.nl
Arcam AB	Bohusgata 13 SE-411 39 Gothenburg Sweden	46-89-856-85-0 Fax 46-31-708-38-38 info@arcam.com	www.arcam.com
ARRK Europe	Development Group Ltd Olympus Park Quedgeley, Gloucester GL2 4NF, UK	44(0) 1452 72770 Fax 44(0) 1452 727755	www.projects@arrkeurope.com
EOS GmbH	Pasinger Strasse 2 D-82152 Planegg/Munich Germany	49-89-89336122 Fax 49-89-856-85-285 info@eos-gmbh.de	www.eos-gmbh.de
Röders GmbH	Scheibenstrasse 6 D-29614 Soltau Germany	49-5191-603-43 Fax 49-5191-603 hsc@roeders.de	www.roeders.com.hk
Swift Technologies	140-144 Station Road March, Cambridgeshire England PE15 8NH	44-1354-650-789 Fax 44-1354-650-799 info@rapid-tech.co.uk	www.swiftech.co.uk

APPENDIX D

RP SOFTWARE DEVELOPERS

Manufacturer	Location	Contacts	Website
Actify, Inc.	60 Spear St. 5th Floor San Fransisco, CA 94105	415-227-3800 Fax 415-227-3802	www.actify.com
Brock Rooney & Associates, Inc.	915 Westwood Birmingham, MI 48009	248-645-0236	www.4dgraphics.net
Delf Spline Systems	Vogelsanglaan 30 3571 ZM Utrecht The Netherlands	313-0-2965-959 Fax 313-0-2962-292	www.deskproto.com
Markam Engineering, GmbH	Fahrenheiststrasse 1 D-28359 Bremen Germany	49 (0)421-2208-336 Fax 49 (0) 421-2208-336	www.marcam.de
Materialise, GmbH	Technologielaan 15 3001 Leuven Belgium	32-16-39-66-11 Fax 32-16-39-66-00 software@materialise.be	www.materialise.com
Matrrialise, Inc., U.S.A.	6111 Jackson Rd. Ann Arbor, MI 48103	734-662-5057 Fax 734-662-7891 info@materilise.com	www.materialise.com
Metris N.V.	Interleuvenlaan 15 D B-3001 Leuven Belgium	321-6740-101 Fax 3216-740-102	www.paraform.com
Solid Concepts, Inc.	28231 Avenue Crocker, Bldg. 10 Valencia, CA 91355	661-257-9300	www.solidconcepts.com

APPENDIX E

RP MATERIAL SUPPLIERS

Manufacturer	Location	Contacts	Website
3D Systems, Inc.	26081 Avenue Hall Valencia, CA 91355	888-337-9786 661-295-5600 Fax 661-295-8404	www.3dsystems.com
American Dye Source, Inc.	555 Morgan Blvd. Baie D'Urfe Quebec, Canada H9X 3T6	514-457-0070 Fax 514-457-0071	www.tuxedoresin.com
BJB Enterprises, Inc.	14791 Franklin Avenue Tustin California 92780	714-734-8450 Fax 714-734-8929 customerserv@bjbenterprises.com	www.bjbenterprises.com
Bolson Material Corporation	8318-106 Street Edmonton Alberta Canada, T6E 4J1	+1-780-669-1587 Fax +1-780-669-1468 support@bolsonmaterials.com	www.bolsonmaterials.com
Cubic Technologies, Inc.	1000 E. Dominguez Street Carson, CA 90746	310-965-0129 Fax 310-965-0141	www.cubictechnologies.com
DSM Somos	Two Penn's Way, Suite 401 New Castle, DE 19720	302-328-5435 Fax 302-328-5693 charles.kaufmann@dsm.com	www.dsmsomos.com
Freeman Supply Company	15147 Don Julian Road City of Industry, CA 91746	800-325-2100 Fax 440-9343-7200	www.freemancompany.com
Objet Geometries Ltd.	Kiryat Weizmann Science Park P.O. Box 2496 Rehovot, 76124 Israel	972-8-9314-313 Fax 972-8-9314-313 info@2objet.com	www.objet.co.il

Solidscape, Inc.	316 Daniel Webster Highway Merrimack, NH 03054-4115	603-429-9700 Fax 603-424-1850 precision@solid-scape.com	www.solid-scape.com
Stratasys Inc.	14950 Martin Drive Eden Prairie, MN 55344-2020	888-480-2548 952-937-3000 Fax 952-937-0070 info@stratasys.com	www.stratasys.com
Vatico, Inc., Renshape Solutions	Unit 29 Keith Road Winnipeg, MB, R3H 0H7	+1-204-694-1660 Fax +1-204-697-1584 India +91-22-5697-5227 Fax +91-22-5697-5237	www.renshape.com
Z Corporation	32 Second Avenue Burlington, MA 01803	781-852-5005 Fax 781-852-5100 sales@zcorp.com	www.zcorp.com

GLOSSARY

3D Printing A low-cost variation of RP that is generally more efficient, user friendly, and less expensive.

ABS acrylonitryl–butadiene–styrene; thermoplastic RP material for Stratasys FDM machine.

CAD (computer-aided design) CAD is the "front-end" for most rapid prototyping systems.

CAE (computer-aided engineering) CAE software allows for engineering analysis that includes determining a design's structural integrity or its heat transfer capacity.

CAM (computer-aided manufacturing) CAM generally refers to a system that uses CAD surface data to drive CNC machines, such as mills and lathes, in order to create parts, molds, and dies.

CNC (computer numerical control) Mill, lathes, and flame cutters are machines that are equipped with CNC capabilities.

Facet Three- or four-sided polygon elements that represents a piece of a 3D polygonal mesh surface of model. Triangular facets are used in STL files.

FDM (fused deposition modeling) In this method, a model created in a CAD program is imported into a software program specifically designed to work with the FDM machine

FFF (freeform fabrication) A more descriptive name for rapid prototyping methods.

IGES (initial graphics exchange specifications) IGES is a standard industry format for exchanging CAD data between systems.

LOM (laminated object manufacturing)   This process generates a part by laser trimming materials to a desired cross section. The material is delivered in sheet form. The cross sections cut from the sheets are laminated into a solid block form using a thermal adhesive coating.

Prototype Tooling   Molds, dies, and other devices used to produce prototypes are sometimes referred to as soft tooling.

RE (reverse engineering)   The science of taking an existing physical model and reproducing its surface geometry in a 3D data file on a CAD system.

RIM (reaction injection molding)   This process uses a simple resin injection system with two pressurized chambers.

RP (rapid prototyping)   A layer-by-layer additive process driven by computer model data joins liquid, powder, or sheet materials to create free-form objects in plastic, wood, ceramic, metal, or composite materials.

RT (rapid tooling)   Tooling that is driven from an RP process—the key to making it rapid. An approach to RT includes producing tooling components, such as mold inserts, directly from an RP process.

Service Bureau   A service bureau is a company that owns and operates one or more rapid prototyping machines and will contract with other companies to use those machines to produce prototypes for them

SLS (Selective Laser Sintering)   First, a part is built by sintering when a laser beam hits a thin layer of powder material. Second the part is built layer by layer. The next layer of powder is deposited by a roller mechanism on top of the previously formed layer. That powder layer is then sintered to the previous layer by the laser.

Solid Model   A 3D CAD model defined using sold modeling techniques. Solid models are preferred over surface models for RP because they define a closed, "water-tight" volume—a requirement of most RP systems.

STL (Stereolithography)   STL file formats are used to convert 3D CAD model data to physical parts using RP systems. The STL format, which is available in binary and ASCII forms, uses triangular facets to approximate the shape of an object.

Surface Model   A 3D CAD model defined by surfaces using a precise mathematical description such as Bezier B-spline surfaces or nonuniform rational B-spline surfaces or nonuniform rational B-spline (NURBS) surfaces. Surface models may or may not represent a closed volume.

Tooling   Molds, dies, and other devices for applications such as plastic injection molding, die casting, and sheet metal stamping.

LIST OF ABBREVIATIONS

2D	two dimensional
3D	three dimensional
3DP	three-dimensional printing
ABS	acrylonitrile–butadiene–styrene
ACES	accurate clear epoxy solid
AIM	ACES injection molding
ANOVA	analysis of variance
ANSI	American National Standards Institute
AQL	acceptable quality levels
ASCII	American Standard Code for Information Interchange
BASS	break away support system
bcc	body-centered cubic
BPM	ballistic particle manufacturing
B-rep	boundary representation
BS	butadiene–styrene
CAD	computer-aided design
CAE	computer-aided engineering
CAM	computer-aided manufacturing
CAT	computed axial tomography
CCD	charged coupled device
CD	computer disk
CIM	computer-integrated manufacturing
CLI	common layer interface
CMB	controlled metal buildup
CMM	coordinate measuring machine
CNC	computer numerical control

CRADA	Cooperative Research and Development Agreement
CSG	constructive solid geometry
CT	computerized tomography
DARPA	Defense Advanced Research Project Agency
DCM	direct composite manufacturing
DMLS	direct metal laser sintering
DNA	deoxyribonucleic acid
DOE	design of experiments
DOF	depth of field
DSP	digital signal processor
DSPC	direct shell production casting
EB	electron beam
EDM	electric discharge machine
EOS	Electrical Optical Systems
EPA	Environmental Protection Agency
fcc	face-centered cubic
FDA	Food and Drug Administration
FDM	fused deposition modeling
FEA	finite element analysis
FEM	finite element method
FFF	freeform fabrication
FMS	flexible manufacturing system
FOV	field of view
GARPA	Global Alliance of Rapid Prototyping
GRIN	gradient index
hcp	hexagonal close packed
HDPE	high-density polyethylene
HIP	hot isostatic pressing
HPGL	Hewlett-Packard graphics language
HR	high resolution
ICP	iterative closest point
IGES	initial graphics exchange specifications
IPT	Institute of Production Technology
JIT	just in time
JPL	Jet Propulsion Laboratory
LAIP	Laboratory to Advance Industrial Prototyping
LAM	laser additive manufacturing
LCC	local composition control
LDPE	low-density polyethylene
LEAF	layer exchange ASCII format
LENS	laser engineered net shaping
LMC	liquid molding compound
LMT	layer manufacturing technologies
LMU	Loyola Marymount University
LOM	laminated object manufacturing

MIMICS	Materialise's Interactive Medical Image Control System
MJM	multijet modeling
MJS	multiphase jet solidification
MRI	magnetic resonance imaging
MSOE	Milwaukee School of Engineering
NASUG	North American Stereolithography User Group
NC	numerical control
NMRI	nuclear magnetic resonance imaging
NSF	National Science Foundation
NURBS	nonuniform rational B-spline
ODM	On Demand Manufacturing
ONR	Office of Naval Research
OVG	Otto van Guericke
PDS	plasma discharge spheroidization
PE	polyethylene
PET	positron emission tomography
PLT	paper lamination technology
PM	powder metallurgy
POM	Precision Optical Manufacturing
PPS	polyphenylene sulfide
PPSF	polyphenylsulfone
PSZ	partial stabilized zirconia
RE	reverse engineering
REU	Research Experience for Undergraduates
RF	radio frequency
RIM	reaction injection molding
RM	rapid manufacturing
rms	root mean square
ROI	region of interest
RP	rapid prototyping
RP&M	rapid prototyping & modeling
RPA/SME	Rapid Prototyping Association of the Society of Manufacturing Engineers
RPC	Rapid Prototyping Center
RPI	rapid prototyping interface
RPM	rapid plaster molding
RPS	rapid prototyping systems
RPT	rapid prototyping technologies
RPTM	rapid prototyping, tooling, and manufacturing
RT	rapid tooling
RTS	rapid tooling system
RTV	room temperature vulcanization
SAE	Society of Automotive Engineers
SAHP	selective adhesive and hot press
SCF	shrinkage compensation factors

SCS	solid creation system
SDM	shape deposition manufacturing
SEM	scanning electron microscope
SFF	solid freeform fabrication
SFM	solid freeform manufacturing
SGC	solid ground curing
SLA	stereolithography apparatus
SLC	stereolithography contour
SLS	selective laser sintering
SML	Stratasys machine language
SOUP	solid object ultraviolet laser plotting
SPI	Sanders Prototype, Inc.
STL	stereolithography file
TS	total shrinkage
TTZ	transformation-toughened zirconia
UOC	ultrasonic object consolidation
UV	ultraviolet

INDEX